From Science to God

also by Peter Russell

The TM Technique
The Brain Book
The Upanishads
The Global Brain Awakens
The Creative Manager
Waking Up in Time
The Consciousness Revolution

From Science to God

The Mystery of Consciousness and the Meaning of Light

PETER RUSSELL

Elf Rock Productions
2375 E. Tropicana Ave., #733
Las Vegas, NV 89119

Cover design by Blue Design, Portland, ME
Cover photo by Joseph Sohm/Visions of America, Ojai, CA
Book design by Julie Donovan, Marin County, CA

ISBN: 1-928586-06-6

Library of Congress Catalog Card Number: 99-61036

PRINTED IN THE UNITED STATES OF AMERICA ON RECYCLED PAPER
10 9 8 7 6 5 4

Contents

Acknowledgments VII

Preface to the Pre-Publication Edition IX

Introduction 1

1. From Science to Consciousness 3

2. The Anomaly of Consciousness 17

3. A Sentient Universe 31

4. The Illusion of Reality 39

5. The Mystery of Light 57

6. The Light of Consciousness 69

7. Consciousness as God 87

8. The Meeting of Science and Spirit 103

9. The Great Awakening 119

Acknowledgments

Many people have been essential to this book. Julie Donovan, my production manager, never ceased in her encouragement, and once I had finished, took the book through all the production stages from copy editing to design, page layout, and printing.

Zorica Gojkovic's help was invaluable. She painstakingly worked through the final drafts with me, smoothing the structure, easing the flow, and making the book a much easier read.

Tinker Lindsay went through the book at many different stages of its development, adding the immeasurable benefit of her own writing experience.

Bocarra Legendre, Christian de Quincey, Cynthia Alves, Dewitt Jones, Karen Malik, and the late Dave Emmer gave me insightful and helpful advice on various versions of the book.

I am deeply grateful to the Fetzer Institute, whose generous grant enabled me to focus my time on writing, and so complete the book much sooner than would otherwise have been possible.

Finally, I would like to thank everyone at the Institute of Noetic Sciences for their continued encouragement and support.

Preface

The first version of this book was a pre-publication edition. With previous books, I made reading copies of the final manuscript for friends, publishers, publicists, and others. With today's technologies, a small print run had become an attractive alternative. I could seed the ideas more widely and get feedback from more people.

As a result of feedback from many readers, I have made some changes in this first trade edition. I have reworked the material concerning the relationship between light and God, and the mystical identity of self and God, to make my meaning clearer. In addition, I have rewritten a large part of the final chapter to reflect my own changing views.

A regularly updated list of recommended reading is available on my website at www.peterrussell.com.

Peter Russell
Sausalito, California
2001

More than anything else, the future of civilization depends on the way the two most powerful forces of history, science and religion, settle into relationship with each other.

Alfred North Whitehead

Introduction

It was the spring of 1996; I had been invited to a small seminar, deep in the California redwoods, to discuss the evolution of consciousness. As I sat there listening to various debates about the nature of mind, recent discoveries in neurochemistry, and theories on the origins of consciousness, I felt increasingly frustrated. I wanted to say, "We've got it all backwards," or words to that effect. But I couldn't express my misgivings in a coherent, well-reasoned manner—which one needs to do in those settings to be taken seriously. So I bit my lip and sat with my frustration.

A few weeks later, on a plane from Los Angeles to San Francisco, I opened an old book I had recently come across. The author, a Dutchman writing in the 1920s, was not saying anything that was new to me, but he reminded me of the processes of perception and the way we construct our experience of reality. My readings in philosophy, particularly the writings of Immanuel Kant, came flooding back; so did my studies in physics on the nature of light, and my explorations into Eastern philosophy and meditation.

Suddenly the root of my frustration became clear. We need more than a new theory of consciousness. We must reconsider some of our fundamental assumptions about the nature of reality. That was the insight that was trying to break through at the seminar. I started scribbling, and by the time the plane landed, the picture was clear. Our whole worldview needed to be turned inside out.

Over the following months, I worked on an essay pulling together the various pieces of a model of reality in which consciousness played a primary role. In the process, I discovered that the implications were even deeper than I had supposed. The new worldview not only changed the way science looked at consciousness, it also led to a new view of spirituality—and, most surprisingly, to a new concept of God.

The seeds sown on that plane flight have now grown into this book. As with any exploration of such profound issues, the ideas are not complete, and may never be complete. They represent my current thinking on the key ingredients of a new worldview, and how consciousness could be the long-awaited bridge between science and spirit.

As much as the book is a journey of ideas that starts with science and arrives at God, it is also my own personal journey from being a physicist with little interest in spiritual matters to an explorer of consciousness who now begins to appreciate what the great spiritual teachings have been saying for thousands of years.

1

From Science to Consciousness

> People travel to wonder at the height of
> mountains, at the huge waves of the sea,
> at the long courses of rivers, at the vast
> compass of the ocean, at the circular
> motion of the stars; and they pass by
> themselves without wondering.
>
> St. Augustine

I have always been a scientist at heart. As a teenager, I delighted in learning how the world works—how sound travels through the air, why metals expand when heated, why bleaches bleach, why acids burn, how plants know when to bloom, how we see color, why a lens bends light, how spinning tops keep their balance, why snowflakes are six-pointed stars, and why the sky is blue.

The more I discovered, the more fascinated I became. At sixteen I was devouring Einstein and marveling at the paradoxical world of quantum physics. I delved into different theories of how the universe began, and pondered the mysteries of space and time. I had a passion for knowing, an insatiable curiosity about the laws and principles that governed the world.

I was equally intrigued by mathematics, sometimes called "the queen and servant" of science. Whether it was the swing of a pendulum, the vibrations of an atom, or the path of an arrow shot into the wind, every physical process had an underlying mathematical expression. The premises of mathematics were so basic, so obvious, so simple, yet from them unfolded rules governing the most complex of phenomena. I remember the exhilaration I felt upon discovering how the same basic equation—one of the simplest and most elegant of all mathematical equations—governs the propagation of light, the vibrations of a violin string, the coiling of a spiral, and the orbits of the planets.

Matter has reached the point of beginning to know itself. . . . [Man is] a star's way of knowing about stars.

George Wald

Numbers, so boring to many, were to me magical. Irrational and imaginary numbers, infinite series, indefinite integrals—I could not get enough of them. I loved the way they all fitted together, like pieces of a cosmic jigsaw puzzle.

Most intriguing of all was how the whole world of mathematics unfolded by the simple application of reason. It seemed to describe a preordained universal truth that transcended matter, time, and space. Mathematics depended on nothing, and yet everything depended on it. If you had asked me then whether there was a God, I would have pointed to mathematics.

The Young Atheist

Conventional religion I had rejected at an early age. I was brought up as a member of the Church of England, but in a somewhat lax fashion. Like many families in our village, we attended Sunday service every few weeks—enough to keep our sins in check and our guilt at bay. That was as far as religion affected me. It was an accepted part of life, but not an important part.

So it was until I entered my teens, when I went through the customary ceremony of confirmation. If the process had lived up to its name, I should then have been confirmed as a member of the church. Nothing could have been further from the truth. If anything was confirmed, it was my skepticism toward religion.

I could accept ideas of not sinning, loving thy neighbor, caring for the sick and other models of Christian behavior, but my mind balked at some of the articles of faith I was expected to accept. On Sundays, the congregation dutifully recited the Nicene Creed, professing their belief in "God, the Father, creator of Heaven and Earth . . . [whose] only begotten son . . . born of the virgin Mary . . . arose from the dead . . . and ascended into Heaven to sit at the right hand of the Father." Such a creed might have been credible at its inception seventeen hundred years ago, but to a budding scientist growing up in the second half of the twentieth century, it was far from believable.

Copernicus had shown that we were not the center of the universe. Astronomers had found no evidence of a heaven up in the sky. Darwin had dispelled the idea that God created the earth and all its living creatures in six days. And biologists had proven virgin birth impossible. Which story should I believe? A text whose only authority was itself, and whose proclamations had little bearing on my everyday reality? Or contemporary science with its empirical approach to truth? At age thirteen, the choice was obvious. I dropped out of conventional religion, and for the rest of my teens my spiritual concerns were reduced to an ongoing debate as to whether I was an atheist or an agnostic.

Psychological Inclinations

I was not, however, a diehard materialist; I did not believe everything could be explained by the physical sciences. By my mid-teens I had developed an interest in the untapped powers of the human mind. Stories of yogis buried alive for days, or lying on beds of nails, intrigued me. I dabbled in so-called out-of-body experiences and experimented with the altered states of consciousness produced by hyperventilating or staring at pulsating lights. I developed my own techniques of meditation, though I did not recognize them as such at the time. I was fascinated by the possible existence of extraterrestrial intelligence; given the trillions of stars in the cosmos, I thought it extremely unlikely that ours was the only planet in the entire universe that had developed conscious life.

I was also making my first ventures into philosophy. My friends and I spent countless hours debating whether or not the mind had an independent existence from the brain. If so, how did mind and brain interact? Or was the mind somehow generated by the brain? When we tired of that, there was always the related question of free will versus determinism. If everything, including the state of our own brains, was predetermined by the laws of physics, was our experience of free will genuine, or just an illusion?

Nevertheless, drawn as I was to questions about the human mind, my overriding interest was still in the physical sciences, and above all, mathematics. Thus, when the time came to pick which subject I was to study at university, the choice was obvious. And when I had to decide which university I should apply to, the choice was again clear. Cambridge was, and probably remains, the best British university at which to study mathematics.

Premonitions of Paradise

My first visit to Cambridge was on the day of my interview—the first part of the admissions procedure.

From afar, the city emerged from flat, wet, green fields like a cultural oasis. As I approached the city center, streets of neatly terraced houses and Edwardian homes gave way to grander university buildings. Architecture from across the centuries—old Norman churches, towering Gothic chapels, ornate Elizabethan halls, Victorian science laboratories, and contemporary edifices of glass and steel—mingled in the sunshine. Within the college walls, carefully manicured lawns covered the courtyards and quadrangles. Heavy oak doors concealed worn stone staircases, leading to the rooms of who-knew-what world-famous professor.

At the heart of the university lay the market square. Unlike many English towns, whose traditional markets have disappeared to be replaced by tasteful cobblestone reminders of the town's heritage, the Cambridge market square was crowded with stalls full of fruit, vegetables, flowers, clothes, books, records, hardware, toys, furniture, and bric-a-brac. Here was a city whose heart remained alive, whose soul had not been trampled by the twentieth century's rush to efficiency and functionality.

As I made my way through the gently winding streets to the college where I was to have my interview, I had that feeling you sometimes get when you meet someone and just know that you are going to be seeing a lot more of each other. I felt sure that I would be coming to live in this exotic seat of learning.

About six weeks later, as I left my house one morning on my way to school, I passed the postman on his rounds. Out of the blue, the thought came that he had a letter for me, and not just any letter; he had a letter offering me a place at Cambridge. There was no reason to expect such a letter. Although my interview had gone reasonably well, I had not yet taken the entrance exam. So I dismissed the thought and continued on my way.

Arriving at school a half hour later, I was told my mother had just phoned with the news that the postman did have a letter from Cambridge, and that I had indeed been offered a place.

Going Up

Nine months later, I was "going up," as they say at Cambridge, to begin my studies. The day after my arrival I had my first meeting with my tutor, a renowned professor of English literature. At Cambridge a tutor has very little to do with a student's academic instruction; that is the concern of one's supervisor. Tutors are *in locus parentis*, a Latin term meaning "in place of parents." Their role is to take care of a student's personal well-being.

"Don't be too serious a student," my tutor advised. "Go to the lectures, yes; do your assignments. But most of all value the people who are here. Your fellow students are the cream of the cream, and the graduates and dons you will be living with are some of the best minds in the country.

"The conversations you have over dinner, or walking by the river in the afternoon, are as important as the lectures you go to in the morning. You are here not just to get a degree, you are also here to mature as a person, to find yourself."

Never could there have been a more favorable time to find oneself. This was Cambridge in the sixties. Centuries-old traditions were crumbling fast. The university had just abolished the rule that undergraduates should wear their academic gowns when about the town at night. No longer was a male student likely to be expelled for being caught with a woman in his room. Students staged the first sit-ins, challenging the establishment to give them a democratic voice in their own education. A banner demanding "Peace in Vietnam" was strung between the spires of King's College chapel—an audacious, if sacrilegious, feat of climbing. There was a sense of hope, the potential for change, for something new.

Peace was in the air, and so was love. Hippies in Afghan coats happily rubbed shoulders with students in tuxedos. White bicycles appeared, belonging to no one but available for anyone to use. Karl Marx, Alan Watts, and Marshall McLuhan were prescribed extracurricular reading. Sergeant Pepper called out across the courtyards, inviting anyone and everyone to sit back and enjoy the show.

The Turning Point

I was exactly where I thought I wanted to be, studying with the best of minds in the best of places. By my third year, Stephen Hawking was my supervisor. Although he had already fallen prey to the motor-neuron disorder known as Lou Gehrig's disease, the illness had not yet taken its full toll. He could still walk with the aid of a cane and speak well enough to be understood.

Sitting with him in his study, half my attention would be on whatever he was explaining—the solution of a particularly difficult set of differential equations, perhaps—but my eye would be caught by hundreds of sheets of paper strewn across his desk. Scrawled on them, in very large handwriting, were equations that I could hardly begin to fathom. Only later did I realize they were probably part of his seminal work on black holes.

On more than one occasion, a spasmodic movement of his arm would send a pile of papers sliding to the floor. I wanted to get down and scoop them up for him, but he always insisted I leave them there. To be doing such groundbreaking work in cosmology was achievement enough. To be doing it with such handicaps was astounding. I felt extremely privileged and very daunted.

Yet, deep inside, something else was stirring.

In mathematics I had reached the point where I could solve Schrödinger's equation for the hydrogen atom. Schrödinger's equation is one of the fundamental equations of quantum physics. Solving it for a single particle such as an electron is fairly straightforward; solving it for two particles—the electron and proton that constitute a hydrogen atom—is more difficult. But once you have the solution, you can begin to make predictions about the behavior of the atom. To me, this was fascinating. Out of pure mathematics emerged functions that described the physics of hydrogen, and to some degree its chemistry.

But now another more intriguing question had aroused my interest. How had hydrogen, the simplest of elements, evolved into creatures such as ourselves, able to reflect upon the immensity of the cosmos, understand its functioning, and even study the mathematics of hydrogen? How had a transparent, odorless gas, become a system that could be aware of itself? In short, how had the universe become conscious?

The most incomprehensible thing about the universe is that it is comprehensible.

Albert Einstein

However hard I studied the physical sciences, they were never going to answer these deeper, more fundamental, questions. I felt increasingly drawn to explore mind and consciousness, and focused less on my mathematical assignments.

My tutor, sensing my distraction, approached me one day to ask how I was doing. I shared with him as best I could my misgivings about my chosen path. His response surprised me: "Either complete your degree in mathematics [I was in my final year] or take the rest of the year off and use it to decide what you really want to study." Then, knowing how hard it would be for me to make such a choice, he added, "I want your decision by noon on Saturday."

Saturday, five minutes before noon, I was still torn between my two options, struggling with feelings of failure and a sense of wasted time, yet knowing I would not be fulfilled continuing with mathematics. In the end I surrendered to my intuition and decided to take the rest of the year off. By late afternoon, I had packed, said a temporary farewell to my friends, and was on my way, with only uncertainty ahead.

The Best of Both Worlds

During the next six months I produced light shows, worked in a jam factory at night, and from time to time pondered my future career.

At first I thought I might study philosophy.

The term *philosophy* originated 2,500 years ago with Pythagoras, best known to most of us for his mathematical explorations. Pythagoras led a remarkable life, even by today's standards. As a teenager he made his way from Greece to Egypt, where he spent ten years training as a temple initiate. His career was interrupted by the Persians, who raided Egypt and took Pythagoras back to Babylon as a slave. Ten years later, his learning and wisdom earned him his freedom; but then, rather than returning to his native Greece, he remained in Babylon for another ten years, studying mathematics in the mystery schools. When he did eventually return home, he established a community in Southern Italy, where he shared with his students much of what he had learned over the years.

Pythagoras was a puzzle to his contemporaries; his life did not fit any conventional style. When asked by a visitor to his community what it was he did, he is said to have replied, "I am simply a lover [*philo*] of wisdom [*sophia*]."

Philosophy at Cambridge had changed considerably from the love of wisdom. Mostly it was the study of past philosophers. Where living philosophers were concerned, logical positivism was the vogue, and I'd had enough of logic by then. None of it had much to do with the questions concerning the nature of consciousness.

The aim isn't to degrade mind to matter, but to upgrade the properties of matter to account for mind, and to tell how from the dust and water of the earth, natural forces conjured a mental system capable of asking why it exists.

Nigel Calder

The only other academic discipline that broached the subject of consciousness was experimental psychology. Whereas clinical psychology involved treating those who are mentally ill, experimental psychology was concerned with the normal functioning of the human brain. It also included learning, memory, the processes behind perception, and how the brain builds up its picture of the world. I decided it was a step in the right direction, and returned to university to study experimental psychology.

The structure of degrees at Cambridge was a little different from most other universities. Degrees were awarded within a particular school, and you could only combine subjects from within that school. Mathematics, for example, came under the School of Mathematics, and could not be combined with philosophy, which came under the School of Moral Sciences. Experimental psychology came under the School of the Natural Sciences. So did theoretical physics. Since they were in the same school, I could combine them into a single degree. Moreover, the curriculum for theoretical physics was essentially the same as that of applied mathematics—in many cases the lectures were identical, and often given by the same professors; only the buildings and course titles were different.

So I found myself able to continue pursuing my interests in mathematics and physics, while at the same time embarking on my exploration of the inner world of consciousness.

2

The Anomaly of Consciousness

A new scientific truth does not triumph
by convincing its opponents and making
them see the light, but rather because its
opponents eventually die.

MAX PLANCK

Today, after thirty years of investigation into the nature
of consciousness, I have come to appreciate how big a problem consciousness is for contemporary science. Science has had remarkable success in explaining the structure and functioning of the material world, but when it comes to the inner world of the mind—to our thoughts, feelings, sensations, intuitions, and dreams—science has very little to say. And when it comes to consciousness itself, science falls curiously silent. There is nothing in physics, chemistry, biology, or any other science that can account for our having an interior world. In a strange way, scientists would be much happier if there were no such thing as consciousness.

David Chalmers, professor of philosophy at the University of Arizona, calls this the "hard problem" of consciousness. The so-called "easy problems" are those concerned with brain function and its correlation with mental phenomena: how, for example, we discriminate, categorize, and react to stimuli; how incoming sensory data are integrated with past experience; how we focus our attention; and what distinguishes wakefulness from sleep.

To say these problems are easy is a relative assessment. Solutions will probably entail years of dedicated and difficult research. Nevertheless, given sufficient time and effort, we expect that these "easy problems" will eventually be solved.

The really hard problem is consciousness itself. Why should the complex processing of information in the brain lead to an inner experience? Why doesn't it all go on in the dark, without any subjective aspect? Why do we have any inner life at all?

I now believe this is not so much a hard problem as an impossible problem—impossible, that is, within the current scientific worldview. Our inability to account for consciousness is the trigger that will, in time, push Western science into what the American philosopher Thomas Kuhn called a "paradigm shift."

Paradigms

The word *paradigm* (derived from the Greek *paradigma,* meaning "pattern") refers to the commonly accepted theories, values, and scientific practices that constitute "normal science" within any particular discipline. A paradigm is a school of thought, a set of assumptions within which a particular science operates. Quantum theory, Newtonian mechanics, chaos theory, Darwin's theory of evolution, and the psychoanalytic model of the unconscious mind are all examples of paradigms.

Over time paradigms change. For nearly two thousand years Plato's theories governed the way people thought about the motion of heavenly bodies. In the seventeenth century Newton's laws of motion became the paradigm. Today, Einstein's theories of relativity are regarded as a more accurate description of how matter moves in space and time. Similar changes in worldview can be found in biology, chemistry, geology, psychology—indeed, in all the sciences.

———————————

All *descriptions of reality are temporary*
hypotheses.

Buddha

———————————

In his seminal book, *The Structure of Scientific Revolutions,* Thomas Kuhn showed that the transition from one paradigm to the next is not smooth. The pressure for change builds over time, but the shift itself is abrupt.

The process begins when the existing paradigm encounters an anomaly—an observation that cannot be explained by the current worldview. Because our assumptions as to how the world works are so deeply ingrained, the anomaly is initially overlooked, or rejected as an error. Or, if it cannot be so easily discarded, attempts are made to incorporate the anomaly within the existing paradigm. This is what happened when medieval astronomers tried to explain the motions of the planets through the sky.

Defending the Paradigm

For more than a thousand years, astronomers had interpreted their observations based on the model formulated by the Greek philosopher Ptolemy, around A.D. 140: The sun, moon, planets, and stars all revolved around the earth in circular orbits.

But there were problems with this model. Although the stars appeared to move smoothly along circular orbits, the planets did not. They wandered among the stars,[1] their orbits wobbled, their speed varied, and they occasionally appeared to reverse direction in what is known as *retrograde* motion. This was an anomaly the existing *geocentric* (i.e., earth-centered) paradigm could not explain.

The solution astronomers came up with was a system of epicycles—the paths traced out by circles that are themselves rolling around larger circles. If the planets moved along epicycles, this would explain some of the strange planetary motions without having to give up the idea of circular motion.

---

[1]The word *planet* comes from the Greek word *planeta* meaning "wanderer."

As more accurate data was collected, it became apparent that simple epicycles were not sufficient to explain all the irregularities. So the medieval astronomers proposed more complex epicycles—circles rolling around circles rolling around circles. When these, too, failed to account for all the observations, they added other modifications and oscillations, making the system yet more cumbersome.

The Copernican Revolution

Kuhn showed that a paradigm starts to shift when some brave soul challenges the assumptions behind the existing worldview and proposes a new model of reality. Often, however, the new model runs so counter to the existing worldview that it is initially rejected, or even ridiculed, by the establishment.

In the early sixteenth century the Polish astronomer Nicolaus Copernicus proposed just such a radically different worldview. The reason the stars appeared to orbit the earth, he suggested, was that the earth itself was moving, spinning on its own axis. The apparent motion of the heavens was an illusion caused by the motion of the observer.

Copernicus not only proposed that the earth was not stationary; he suggested it was not even at the center of the universe. He found that the anomalous movements of the planets could be explained if they were assumed to be orbiting the sun rather than the earth. From this came his most heretical conclusion: The earth itself was just another planet going around the sun.[2]

It is easy for us, born into a world in which the *heliocentric* (i.e., sun-centered) model is the accepted truth, to overlook just how radical a proposal this was. The earth's central position was not only an article of faith upon which everyone agreed, it was also confirmed by personal experience. One had only to look up to see the sun and stars moving across the sky, while the earth clearly remained as still as could be. To suggest that the earth moved was ludicrous.

Every truth passes through three stages
before it is recognized.
In the first, it is ridiculed.
In the second, it is opposed.
In the third, it is regarded as self-evident.
                          Arthur Schopenhauer

Copernicus was a clergyman and knew his theory not only went against common sense but also challenged the church's view of reality. So, for thirty years, he kept his ideas

---

[2]This was not a totally new theory. In 270 B.C. a little-known Greek philosopher, Aristarchus, advanced the idea that the earth and the other planets moved around the sun. If his views had held sway—rather than those of Plato and Ptolemy—history might have taken a very different course.

to himself. Only as he neared death and felt he did not want to take this important knowledge with him to the grave did Copernicus finally decide to publish. The first copy of his little book, *On the Revolutions of the Celestial Spheres,* arrived in his hands on the day he died.

Copernicus's fears of repression turned out to be well founded. The Vatican disapproved, later placing his work on the papal index of forbidden books, and it remained ignored and forgotten for nearly seventy years.

Completing the Paradigm Shift

In 1609 the Italian scientist Galileo Galilei, using his newly invented telescope, found convincing evidence in favor of Copernicus's ideas. He saw that Venus, like the moon, moved through phases—sometimes only half, or just a crescent, of the planet would be illuminated—which showed that Venus did indeed circle the sun. Galileo also discovered moons orbiting Jupiter, further dispelling the idea that everything circled the earth.

After Galileo published his findings, he was contacted by the Pope, who demanded Galileo retract his heretical ideas. A few years earlier, the philosopher Giordano Bruno had been burned at the stake in Rome for supporting Copernicus's model, so Galileo wisely accorded with the Pope's demands.

But Galileo was not happy that so important a truth should remain suppressed. In 1632 he published *Dialogue,* a brilliantly composed book in which he again defended the Copernican theory. Once more the Vatican demanded a retraction. Galileo was forced to "abjure, curse, and detest" the view that the earth moved around the sun, and was condemned to house arrest for the remainder of his life.

To assert that the earth revolves around the sun is as erroneous as to claim that Jesus was not born of a virgin.

Cardinal Bellarmine
(during the trial of Galileo)

Meanwhile, a German mathematician, Johannes Kepler, was solving another piece of the planetary puzzle. Kepler had had the good fortune to study under Tycho Brahe, a Danish astronomer who had accumulated a vast inventory of accurate astronomical data. These clearly showed that even if the planets were orbiting the sun, they were not following circular orbits. After pondering the data for many years, Kepler found that he could explain all the irregularities in the planets' movements if he assumed they followed elliptical orbits. But as to why this should be, he had no idea.

The answer came seventy years later when the English mathematician Isaac Newton realized that heavenly bodies are governed by exactly the same laws as earthly objects— the force that causes an apple to fall is the same force that holds the moon in its orbit around the earth. Working out the resulting equations of motion, he proved that any orbiting body would move in an ellipse, just as Kepler had discovered.

With this final piece of the puzzle, the revolution was complete. Copernicus had provided the key idea, but it had taken several other equally significant breakthroughs, involving people from five countries, spread over 150 years, to put the sun firmly at the center of things and irrevocably shift the way people viewed their world.[3]

The Metaparadigm

The process by which the geocentric worldview changed to a heliocentric one is a classic example of a paradigm shift occurring in a particular area of science. Yet Kuhn's model need not be limited to individual scientific disciplines. I believe the model can, and should, be taken a step further and applied to the worldview of Western science as a whole.

All our scientific paradigms are based on the assumption that the physical world is the real world, and that space, time, matter, and energy are the fundamental components of reality. When we fully understand the functioning of the physical world, we will, it is believed, be able to explain everything in the cosmos.

This is the belief upon which all our scientific paradigms are based. It is, therefore, more than just another paradigm; it is a *metaparadigm*—the paradigm behind the paradigms.

---

[3]However, it was not until 1992 that the Vatican formally apologized for its treatment of Galileo.

So successful has this metaparadigm been at explaining just about every phenomenon we encounter in the material world, it is seldom, if ever, questioned. It is only when we turn to the nonmaterial world of the mind that this worldview begins to exhibit weaknesses.

Nothing in Western science predicts that any living creature should be conscious. It is easier to explain how hydrogen evolved into other elements, how they combined to form molecules and then simple living cells, and how these evolved into complex beings such as ourselves than it is to explain why we should ever have a single inner experience.

Scientists are in the strange position of being confronted daily by the indisputable fact of their own consciousness, yet with no way of explaining it.

Christian de Quincey

The problem is, in essence, one of *type*. When elementary particles combine to form atoms, and those atoms combine to form molecules, they are forming entities of the same type—they are all physical phenomena. The same is true of a simple cell. DNA, proteins, and amino acids are of the same basic type as atoms. Even the human brain, unfathomably complex as it may be, is still of the same essential type.

Consciousness, however, is of a fundamentally different type. Consciousness is not composed of matter. And matter, we assume, does not possess consciousness.

We may not be able to account for consciousness, yet the fact that we are conscious is one thing of which we are absolutely certain. This realization was one of René Descartes's great contributions to Western philosophy, some three hundred and fifty years ago. Like many philosophers before and since, Descartes was looking for absolute truth. To this end, he created his method of doubt. Anything that was open to doubt, he argued, could not be the absolute truth.

Descartes found that he could doubt any theory or philosophy. He could doubt what anybody said. He could doubt what his eyes showed him of the world. He could doubt his own thoughts and feelings. He could even doubt that he had a body. But the one thing he could not doubt was that he was doubting. This revealed one certainty: he was thinking. If he was thinking, he had to be an experiencing being. As he put it in Latin, *Cogito, ergo sum*—"I think, therefore I am."

This is the paradox of consciousness. Its existence is undeniable, yet it remains totally inexplicable. For the materialist metaparadigm, consciousness is one big anomaly.

Defending the Metaparadigm

As Kuhn showed, the first reaction to an anomaly is to ignore it. This is what most scientists have done with consciousness, and for what seemed good reasons.

First, consciousness cannot be observed in the way that material objects can. It cannot be weighed, measured, or otherwise pinned down. Second, scientists have sought to arrive at universal objective truths, independent of any particular observer's viewpoint or state of mind. To this end they have deliberately avoided subjective considerations. And third, they felt there was no need; the functioning of the universe could be explained without having to explore the troublesome subject of consciousness.

But developments in several fields have now shown that consciousness cannot be quite so easily sidelined. Quantum physics, for example, suggests that, at the atomic level, the act of observation affects the reality that is observed. In medicine, a person's state of mind can have significant effects on the body's ability to heal itself. As neurophysiologists deepen their understanding of brain function and its correlation with mental phenomena, the nature of subjective experience again raises its head.

As a result of these and other developments, a growing number of scientists and philosophers are now trying to explain how consciousness arises. Some believe that a deeper understanding of brain chemistry will provide the answers; perhaps consciousness resides in the action of neuropeptides. Others look to quantum physics. The minute microtubules found inside nerve cells could create quantum effects that might somehow contribute to consciousness. Some explore computing theory and believe that consciousness emerges from the complexity of the brain's processing. Others find sources of hope in chaos theory.

Yet whatever idea is put forward, one thorny question remains unanswered: How can something as immaterial as consciousness ever arise from something as unconscious as matter?

The continued failure of these approaches to make any appreciable headway into solving this problem suggests they may all be on the wrong track. They are all based on the assumption that consciousness emerges from, or is dependent upon, the physical world of space, time, and matter. In one way or another, they are attempting to accommodate the anomaly of consciousness within a worldview that is intrinsically materialist. As happened with the medieval astronomers who kept adding more and more epicycles to explain the anomalous motions of the planets, the underlying assumptions are seldom, if ever, questioned.

I now believe that rather than trying to explain consciousness in terms of the material world, we should be developing a new worldview in which consciousness is a fundamental component of reality. The key ingredients for this new metaparadigm are already in place. We need not wait for any new discoveries. All we need do is put various pieces of our existing knowledge together and explore the new picture of reality that emerges.

3

A Sentient Universe

> ... a nature found within all creatures
> but not restricted to them; outside all
> creatures, but not excluded from them.
>
> *THE CLOUD OF UNKNOWING*

What is consciousness? The word is not easy to define, partly because we use it to cover a variety of meanings. We might say an awake person has consciousness, whereas someone who is asleep does not. Or, someone could be awake, but so absorbed in their thoughts that they have little consciousness of the world around them. We speak of having a political, social, or ecological consciousness. And we may say that human beings have consciousness while other creatures do not, meaning that humans think and are self-aware.

The way I shall be using the word *consciousness* is not in reference to a particular state of consciousness, or a particular way of thinking, but to the faculty of consciousness—

the capacity for inner experience, whatever the nature or degree of the experience.

For every psychological term in English there are four in Greek and forty in Sanskrit.

A. K. Coomaraswamy

The faculty of consciousness can be likened to the light from a film projector. The projector shines light onto a screen, modifying the light so as to produce any one of an infinity of images. These images are like the perceptions, sensations, dreams, memories, thoughts, and feelings that we experience—what I call the "forms of consciousness." The light itself, without which no images would be possible, corresponds to the faculty of consciousness.

We know all the images on the screen are composed of this light, but we are not usually aware of the light itself; our attention is caught up in the images that appear and the stories they tell. In much the same way, we know we are conscious, but we are usually aware only of the many different perceptions, thoughts, and feelings that appear in the mind. We are seldom aware of consciousness itself.

Consciousness in All

The faculty of consciousness is not limited to human beings. A dog may not be aware of all the things of which we are aware. It does not think or reason as humans do, and it probably does not have the same degree of self-awareness, but this does not mean that a dog does not have an inner world of experience.

When I observe a dog, I infer that it has its own mental picture of the world, full of sounds, colors, smells, and sensations. It appears to recognize people and places, much as we might. A dog may at times show fear, and at other times excitement. Asleep, it can appear to dream, feet and toes twitching as if on the scent of some fantasy rabbit. When a dog yelps or whines we assume it is feeling pain—indeed, if we didn't believe that dogs felt pain, we wouldn't bother giving them anesthetics before an operation.

If dogs possess consciousness, then so do cats, horses, deer, dolphins, whales, and other mammals. They may not be self-conscious as we are, but they are not devoid of inner experience. The same is true of birds; some parrots, for example, seem as aware as dogs. And if birds are sentient beings, then so, I assume, are other vertebrates—alligators, snakes, frogs, salmon, sharks. However different their experiences may be, they all share the faculty of consciousness.

The same argument applies to creatures further down the evolutionary tree. The nervous systems of insects are not nearly as complex as ours, and insects probably do not have as rich an experience of the world as we do, but I see no reason to doubt that they have some kind of inner experience.

Where do we draw the line? We usually assume that some kind of brain or nervous system is necessary before consciousness can come into being. From the perspective of the materialist metaparadigm, this is a reasonable assumption. If consciousness arises from processes in the material world, then those processes need to occur somewhere, and the obvious candidate is the nervous system.

However, we then come up against the inherent difficulty of the materialist metaparadigm. Whether we are considering a human brain with its tens of billions of cells, or a nematode worm with a hundred or so neurons, the problem is the same: How can any purely material process ever give rise to consciousness?

Panpsychism

The underlying assumption of the current metaparadigm is that matter is insentient. The alternative is that the faculty of consciousness is a fundamental quality of nature. Consciousness does not arise from some particular arrangement of nerve cells or processes going on between them, or from any other physical features; it is always present.

If the faculty of consciousness is always present, then the relationship between consciousness and nervous systems needs to be rethought. Rather than creating consciousness, nervous systems may be amplifiers of consciousness, increasing the richness and quality of experience. In the analogy of a film projector, having a nervous system is like having a lens in the projector. Without the lens there is still light on the screen, but the images are much less sharp.

In philosophical circles the idea that consciousness is in everything is called *panpsychism,* from the Greek *pan,* meaning all, and *psyche,* meaning soul or mind. Unfortunately, the words soul and mind suggest that simple life forms may possess qualities of consciousness found in human beings. To avoid this misunderstanding some contemporary philosophers use the term *panexperientialism*—everything has experience.

Whatever name this position is given, its basic tenet is that the capacity for inner experience could not evolve or emerge out of entirely insentient, non-experiencing matter. Experience can only come from that which already has experience. Therefore the faculty of consciousness must be present all the way down the evolutionary tree.[1]

We know that plants are sensitive to many aspects of their environment—length of daylight, temperature, humidity, atmospheric chemistry. Even some single-celled organisms are sensitive to physical vibration, light, and heat. Who is to say they do not have a corresponding glimmer of awareness? I am not implying they perceive as we do, or that they have thoughts or feelings, only that they possess the faculty of consciousness; there is a faint trace of experience. It may be a billionth of the richness and intensity of our own experience, but it is still there.

According to this view, there is nowhere we can draw a line between conscious and nonconscious entities; there is

---

[1]A much fuller treatment of the arguments for and against pan-psychism and panexperientialism can be found in Christian de Quincey's excellent article, "Consciousness All the Way Down?" *Journal of Consciousness Studies* 1, no. 2 (1994): 217–229.

a trace of experience, however slight, in viruses, molecules, atoms, and even elementary particles.

Some argue this implies that rocks perceive the world around them, perhaps have thoughts and feelings, and enjoy an inner mental life similar to human beings. This is clearly an absurd suggestion, and not one that was ever intended. If a bacterium's experience is a billionth of the richness and intensity of a human being's, the degree of experience in the crystals of a rock might be a billion times dimmer still. They would possess none of the qualities of human consciousness—just the faintest possible glimmer of experience.

The Evolution of Consciousness

If the faculty of consciousness is universal, then conscious-ness is not something that emerged with human beings, or with vertebrates, or at any particular stage of biological evo-lution. What emerged over the course of evolution was not the *faculty* of consciousness, but the various qualities and dimensions of conscious experience—the *forms* of con-sciousness.

Bacteria and algae, the earliest living organisms, had no sensory organs and detected only the most general charac-teristics and changes in their environment. Their experi-ence might be likened to an extremely dim, almost imperceptible hint of light on an otherwise dark screen— virtually nothing compared to the complexity and detail of human experience.

With the evolution of multicellular organisms came the emergence of specific senses. Some cells specialized in sensing light, others in sensing vibration, pressure, or changes in chemistry. Working together, such cells formed sensory organs, increasing the detail and quality of the information available to the organism—and enhancing the quality of consciousness.

In order to process this additional information and distribute it to other parts of the organism, nervous systems evolved. As the flow of information became more complex, central processing systems developed, integrating the different sensory modalities into a single picture of the world.

As brains grew in complexity, new features were added to the image appearing in consciousness. With mammals the limbic system appeared, an area of the brain associated with basic feelings such as fear, arousal, and emotional bonding. Over time, the mammalian brain grew yet more complex, developing a new structure around it, the cerebral cortex. With this came better memory, focused attention, greater intention, and imagination.

The picture appearing in consciousness had by now reached the richness of detail and diversity of qualities that we associate with our own experience. But this is not the end of the story. In human beings another new capacity emerged: speech. And with it, the evolution of consciousness took a huge leap forward.

For the first time, we could use words to communicate experiences with each other. Our awareness of the world was no longer limited to what our senses told us; we could know of events occurring in other places and at other times. We could learn from each other's experiences, and so begin to accumulate a collective body of knowledge about the world.

Most significantly, we began to use language internally. Hearing words in our minds without actually saying them allowed us to talk to ourselves silently in our minds. An entirely new dimension had been added to our consciousness: verbal thought. We could form concepts, entertain ideas, appreciate patterns in events, apply reason, and begin to understand the universe in which we found ourselves.

Then came the most important leap of all. Not only could we reflect upon the nature of the world around us, we could also reflect upon thinking itself. We became self-aware—aware of our own awareness. This opened the door to a whole new arena of development. We became a species that could explore the inner world of the mind and, ultimately, delve into the nature of consciousness itself.

4

The Illusion of Reality

All that we see or seem
is but a dream within a dream.

EDGAR ALLEN POE

The faculty of consciousness is one thing we all share, but what goes on in our consciousness, the *forms* that consciousness takes on, varies widely. This is our personal reality, the reality we each know and experience. Invariably we confuse this personal reality with physical reality, believing ourselves to be in direct contact with the world "out there." But the colors and sounds we experience are not really "out there"; they are all images in the mind, pictures of reality we have constructed. This one fact leads to a radical re-thinking of the relationship between consciousness and reality.

The idea that we never experience the physical world directly has intrigued many philosophers. Most notable was the eighteenth-century German philosopher Immanuel

Kant, who drew a clear distinction between the forms that appear in the mind—what he called the *phenomenon* (a Greek word meaning "that which appears to be")—and the world that gives rise to this perception, which he called the *noumenon* (meaning "that which is apprehended"). All we know, Kant insisted, is the phenomenon. The noumenon, the "thing-in-itself," remains forever beyond our knowing.

A century earlier, the British philosopher John Locke had argued that all knowledge is based on perception caused by external objects acting on the senses. Locke thought perception was passive—the mind simply reflecting the images received by the senses—but Kant proposed that the mind is an active participant in the process, continually shaping our experience of the world. Reality, he believed, is something we each construct for ourselves.

As to the ultimate things we can know
nothing, and only when we admit this do
we return to equilibrium.

Carl Jung

Unlike some of his predecessors, Kant was not suggesting that this reality is the *only* reality. Irish theologian Bishop Berkeley had argued that we know only our perceptions. He then concluded that nothing exists apart from our perceptions, which forced him into the difficult position of having to explain what happened to the world when no one was perceiving it. Kant held that there *is* an underlying reality, but we never know it directly. All we can ever know is how it appears in our minds.

The Image in the Mind

Remarkably, Kant came to these conclusions without any of our current scientific knowledge, or any understanding of the physiology of perception. Today we know much more about how the brain constructs its picture of reality.

When I look at a tree, light reflected from the tree forms an image of the tree on the retina of my eye. Photosensitive cells in the retina discharge electrons, triggering electrochemical impulses that travel down the optic nerve to the visual cortex of the brain. There the data undergoes complex processing that reveals shapes, patterns, colors, and movements. The brain then integrates this information into a coherent whole, creating its own reconstruction of the external world. Finally, an image of the tree appears in my consciousness. Just how my neural activity gives rise to a conscious experience is the "hard problem" mentioned earlier. Though we have no idea how an image appears in the mind, it does happen. I have the conscious experience of seeing a tree.

Similar activities take place with the other senses. A vibrating violin string creates pressure waves in the air. These waves stimulate minute hairs in the inner ear, which send electrical impulses on to the brain. As with vision, the raw data is then analyzed and integrated, culminating in the experience of hearing music.

Chemical molecules emanating from the skin of an apple trigger receptors in the nose, leading to the experience of smelling an apple. Cells in the skin send messages to the brain that lead to experiences of touch, pressure, texture, and warmth.

In short, all that I perceive—everything I see, hear, taste, touch, and smell—has been reconstructed from sensory data. I think I am perceiving the world around me, but all that I am directly aware of are the colors, shapes, sounds, and smells that appear in the mind.

---

*Every man's world picture is and always
remains a construct of his mind, and cannot
be proved to have any other existence.*

Erwin Schrödinger

Our perception of the world has the very convincing appearance of being "out there" around us, but it is no more "out there" than are our nightly dreams. In our dreams we are aware of sights, sounds, and sensations happening around us. We are aware of our bodies. We think and reason. We feel fear, anger, pleasure, and love. We experience other people as separate individuals, speaking and interacting with us. The dream appears to be happening "out there" in the world around us. Only when we awaken do we realize that it was all just a dream—a creation in the mind.

When we say, "It was all just a dream," we are referring to the fact that the experience was not based on physical reality. It was created from memories, hopes, fears, and other factors. In the waking state, our image of the world is based on sensory information drawn from our physical surroundings. This gives our waking experience a consistency and sense of reality not found in dreams. But the truth is, our waking reality is as much a creation of our minds as are our dreams.[1]

I have given everything I see . . .
all the meaning it has for me.
                    A Course in Miracles

The idea that reality is a creation of the mind seems to run counter to common sense. Right now you are aware of the pages in front of you, various objects around you, sensations in your own body, and sounds in the air. Even though you may understand it is all a reconstruction of reality, it still appears that you are having a direct perception of the physical world. And I am not suggesting you should try to see it otherwise. What is important for

---

[1]This is not to suggest that we create physical reality. Some people believe our thinking or attitude can have a direct impact on the state of the physical world. Whether or not this is possible is an open question. Here I refer only to creating our personal experience of reality.

now is the understanding that all experience is an image of reality created in the mind.[2]

Cracks in Reality

Our impression that we are perceiving the world directly is usually quite convincing. Occasionally, however, we may come across phenomena that reveal cracks in our construction of reality. Visual illusions are a good example. They usually occur because the brain misinterprets the sensory data and constructs an image of reality that is either misleading or inconsistent.

A simple example is demonstrated by the illustration below. This drawing of a cube is something we have all seen many times, but is it a cube seen from above, or a cube seen from below?

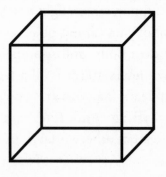

---

[2]The word "image" here means more than just a visual image. The sounds we hear are auditory images. The sensations in our body produce a body image. Taste and smell likewise produce their own kinds of images in the mind.

Most people's first response is "from above," probably because we are used to seeing rectangular corners from above: tables, boxes, TV sets, computers. Less often do we view such objects from below. But if you put your attention on the top line and bring it forward in your mind's eye, you can change your perception and turn it into a cube seen from a different perspective.

The most intriguing aspect of this illustration is not that you can see it in two different ways, but that, whichever way you see it, you see a three-dimensional cube. You are actually seeing twelve lines on a flat sheet of paper, yet your experience is of an object with depth. This depth may appear very real, but it is actually an interpretation added by your brain.

Maya

There are, therefore, two realities: the physical reality—whatever is actually "out there" stimulating our senses—and the personal reality that we each experience, the reconstruction of the world that appears in our minds. And both are very real.

Some people claim that our subjective reality is an illusion, but that is misleading. It may all be a creation of the mind, but it is nonetheless real—the only reality we ever know.

The illusion comes when we confuse the reality we experience with the physical reality, the thing-in-itself. The Vedantic philosophers of ancient India spoke of this confusion as *maya*. Often translated as "illusion" (a false perception of the world), maya is better interpreted as "delusion" (a false belief about the world). We suffer a delusion when we believe the images in our minds *are* the external world. We deceive ourselves when we think that the tree we see is the tree itself.

Things are not what they seem to be,
nor are they otherwise.

Lankavatara Sutra

Our assumption that we are directly interacting with physical reality closely parallels the way we respond to the image on a computer screen. Moving a computer's mouse appears to move the cursor around the screen. In reality, the mouse is sending a stream of data to the central processor, which calculates a new position for the cursor and then updates the image on the screen. In early computers there was a noticeable delay between issuing a command and seeing the effects on the screen. Today computers are so fast they can recalculate the image on a screen in a fraction of a second, and there is no visible delay between the movement of the mouse and the cursor on the screen. We experience moving the cursor across the screen.

Our experience of daily life is similar. When I kick a stone, my intention to move my foot is communicated to my body, and my foot in the physical world moves to meet the physical stone. But I do not experience the interaction directly. The brain receives the information sent back by the eyes and body and updates my image of reality appropriately. As with a computer, there is a small delay between the event in the physical world and my experience of that event. It takes the brain about a fifth of a second to process the sensory information and construct the corresponding picture of reality. Our awareness of reality is about a fifth of a second behind physical reality, but we never notice the lag because the brain cleverly compensates for the delay, leaving us with the impression that we are interacting directly with the physical world.

The Unknowable Reality

If all that we ever know are the sensory images that appear in our minds, how can we be sure there is a physical reality behind our perceptions? Is it not just an assumption? My answer is: Yes, it is an assumption; nevertheless, it seems a most plausible one.

First, there are definite constraints on our experience. For example, we cannot walk through walls. If we try to, we suffer predictable consequences. Nor can we, when awake, float through the air or walk upon water.

Second, our experiences generally follow well-defined laws and principles. Balls thrown through the air follow precisely defined paths. Cups of coffee cool at similar rates. The sun rises on time.

Third, this predictability is consistent. We all experience similar patterns. The simplest way, by far, of accounting for these constraints and for their consistency is to assume that there is indeed a physical reality. We may not know it directly, but we believe it is there.

To reveal the nature of this underlying reality has been the goal of much scientific endeavor. Over the years scientists have elucidated many of the laws and principles that govern its behavior. Yet curiously, the more deeply they have delved into its true nature, the more they discover that physical reality is nothing like we imagined it to be.

This should not be too surprising. If all we can imagine are the forms and qualities that appear in consciousness, then these are unlikely to be appropriate models for describing the underlying physical reality.

For two thousand years atoms were believed to be tiny solid balls—a model clearly drawn from everyday experience. As physicists discovered that atoms were composed of more elementary, subatomic particles (electrons, protons, neutrons) the model shifted to one of a central nucleus surrounded by orbiting electrons—again, a model based on experience.

An atom may be small, a mere billionth of an inch across, but subatomic particles are a hundred thousand times smaller still. Imagine the nucleus of an atom magnified to the size of a grain of rice. The whole atom would then be the size of a football stadium, and the electrons would be other grains of rice flying round the stands. As the early twentieth-century British physicist Sir Arthur Eddington put it, "Matter is mostly ghostly empty space." To be more precise, it is 99.9999999% empty space.

If physical reality is mostly empty space, why does the world seem so substantial and unyielding? Why doesn't the

99.9...% empty space of my hand simply pass straight through the 99.9...% empty space of the table it is resting on? The simplest way of explaining this is that the electrons spin so fast around the nucleus, they create an impenetrable shell through which other particles cannot normally pass. Picture a person swinging a weight around herself on a piece of string; you can never get close enough to touch her, because the circling weight keeps you at bay. In a similar way, when two atoms meet, their electronic orbits stop them from passing through each other, and they behave as if they were solid balls.

Matter is not made of matter.
                                        Hans-Peter Dürr

With the development of quantum theory, physicists have found that even subatomic particles are far from solid. In fact, they are nothing like matter as we know it. They cannot be pinned down and measured precisely. Much of the time they seem more like waves than particles. They are like fuzzy clouds of potential existence, with no definite location. Whatever matter is, it has little, if any, substance.

Seeing What Isn't There

The image of the world that appears in the mind is quite different from the actual physical world, and in two complementary ways.

On the one hand, our image of reality is *more than* physical reality in that it contains many qualities not present in physical reality. Consider our experience of the color green, for example. In the physical world there is light of various frequencies, but the light itself is not green, nor are the electrical impulses that are transmitted from the eye to the brain. No color exists there. The green we see is a quality created in consciousness. It exists only as a subjective experience in the mind.

The same is true of sound. When Bishop Berkeley argued that nothing exists apart from our perceptions, a vigorous debate ensued as to whether a falling tree made a sound if no one was there to hear it. At that time nothing was known of how sound was transmitted through the air, or how the ear and brain functioned. Today we know much more about the processes involved, and the answer is clearly "No." There is no *sound* in the physical reality, simply pressure waves in the air. Sound exists only as an experience in the mind of a perceiver—whether that perceiver is a human being, a deer, a bird, or an ant.

On the other hand, our image of reality is *less than* physical reality in that there are many aspects of the external world we never experience.

Our eyes, for example, are sensitive only to light in the narrow frequency range from 430,000 to 750,000 gigahertz (a gigahertz is a billion cycles per second). At lower frequencies is *infrared* (below red) radiation, and lower still are microwaves and radio waves. At higher frequencies we find *ultraviolet* (above violet) rays, and beyond them X-rays and gamma rays. Our eyes detect none of these other frequencies, and our visual image of reality represents but a tiny fraction of what is there.

The same holds true of the other senses. What we hear, smell, and taste is but a limited sample of physical reality. Furthermore, there are aspects of the physical world, such as magnetic fields and electric charge, that have little, if any, impact on our experience.

Human beings may not be able to sense these other facets of reality, but some creatures can. Dogs, for example, detect much higher frequencies of sound than we do, and their sense of smell is estimated to be a million times more sensitive than ours. If we could put ourselves in a dog's mind we would find ourselves in a different world. Imagine what it might be like to be able to detect the scent of a person hours after they have passed by, and to be able to follow that scent, distinguishing it from hundreds of others, for many miles.

We know that there is not one space and one time only, but that there are as many spaces and times as there are subjects.
                    Jakob von Uexküll

We can fairly easily imagine the reality of a dog, since its sensory perception is an extension of ours. But the reality of a dolphin is much harder to picture. With their highly developed echolocation abilities, dolphins experience qualities of which most of us know nothing.[3] When a dolphin perceives me with its sonar, it does not perceive a solid body. Its sonar image is more like the ultrasound scans used to monitor a fetus during pregnancy. A dolphin can sense the shapes and movements of my internal organs. The beating of my heart, the churning of my stomach and the state of my muscles are all visible to the dolphin mind. It sees my inner reactions as clearly as I see the frown on a person's face.

Other species experience qualities of which we know nothing. Most snakes have organs sensitive to the infrared range of the electromagnetic spectrum, and so "see" the heat emitted by their prey. Bees see in the ultraviolet range, and are sensitive to the polarization of light. Sharks, eels, and other fish can detect minute changes in electrical fields. The realities that they construct contain qualities totally unknown to human experience.

There is no creature who perceives all of what is and what happens.

                    Judith and Herbert Kohl

Ultimately, there are as many different ways of perceiving the world as there are species of life in the universe. What we take to be reality is simply the particular way the human mind sees and interprets the physical world.

---

[3]Some blind people possess an echolocation ability that may give rise to experiences of a similar, though less developed, nature.

The New Copernican Revolution

Immanuel Kant believed his insights into the nature of perception, and the distinction between physical reality and the reality we each experience, would be the basis for "a Copernican revolution in philosophy." Now, two hundred years later, it seems he may have been close to the mark. In the Copernican revolution, the key insight was the realization that the earth was spinning through space. Kant's distinction between the two realities is likewise the key insight which opens the door to a new metaparadigm.

In both cases the key insight defied common sense. In Copernicus's time it seemed absolutely obvious that the earth was still. Today it seems equally obvious that we are perceiving the physical world directly. Even when we intellectually accept the fact that our entire world of experience is a construction within the mind, as eventually we must, we still see the world "out there," around us.

We may always see it this way. Even now, five centuries after Copernicus, we still *see* the sun going down, even though we know that it is really the earth spinning round.

However, it is possible to see it the other way. All you need do is go somewhere where you have a good view of the horizon. Then, rather than thinking of yourself as stationary, see yourself standing on this huge ball of rock we call Earth, which is slowly turning in space from west to east. As it turns, new parts of the sky come into view in the east while others disappear from view in the west. Now, instead of seeing the sun setting, you can see the horizon moving up and hiding it. In a similar way, the full moon "rises" as the opposite horizon moves down, opening up new vistas. Changing your perception in this way, the Copernican shift becomes an experienced reality.

It is much more difficult, however, to do a similar exercise with our perception of the world around us. Try as I may, I cannot experience the fact that it is all an image within my mind. However, this does not mean it is impossible to see things differently. Some spiritual adepts, who have made deep personal investigations into the nature of consciousness, claim to have achieved this new perception.

Perhaps one of the clearest and most succinct descriptions of this alternative mode of consciousness comes from the contemporary Indian teacher Sri Nisargadatta Maharaj, who said, in describing his own spiritual awakening:

> You realize beyond all trace of doubt that the world is in you, and not you in the world.

Swami Muktananda, another contemporary sage, said:

> You are the entire universe.
> You are in all, and all is in you.
> Sun, moon, and stars revolve within you.

And the *Ashtavakra Gita,* a highly venerated Indian text, states:

> The Universe produced phenomenally in me, is pervaded by me. . . . From me the world is born, in me it exists, in me it dissolves.

These people appear to have awakened from the dream of maya—the delusion that we are directly perceiving the physical world. They know as a direct personal experience, not just as a theoretical idea, that their entire world is a manifestation within the mind. These are the ones—the enlightened ones, we sometimes call them—who have personally made the shift to a new metaparadigm.

Turning Reality Inside Out

In much the same way as Copernicus's insight turned our model of the cosmos inside out, the distinction between the physical world and our experience of the world turns the relationship of consciousness and the material world inside out. In the current metaparadigm, consciousness is assumed to emerge from the world of space, time, and matter. In the new metaparadigm, everything we know manifests from consciousness.

Matter is derived from mind, and not mind from matter.
      The Tibetan Book of the Great Liberation

We think the world we see around us is composed of matter. As far as the actual physical reality is concerned, this may be so—uncertain though we may be as to the ultimate nature of this matter. But the world we perceive around us is not the physical world. The world we actually know is the world that takes form in our minds; this world is not made of matter, but of mindstuff. Everything we know, perceive, and imagine, every color, sound, sensation, thought, and feeling, is a form that consciousness has taken on. As far as this world is concerned, everything is structured in consciousness.

Kant argued that this was even true of space and time. To us, the reality of space and time seems undeniable. They appear to be fundamental dimensions of the physical world, entirely independent of our consciousness. This, said Kant, is because we cannot see the world in any other way. The human mind is so constituted that it is forced to construct its experience within the framework of space and time. Space and time are not, however, fundamental dimensions of the underlying reality. They are fundamental dimensions of consciousness.

It was an astonishing claim at the time—and probably still seems astonishing to many of us today—but contemporary physics now lends weight to this extraordinary idea.

5

The Mystery of Light

For the rest of my life I want to reflect on
what light is.

ALBERT EINSTEIN

My decision to study theoretical physics along with experimental psychology was most fortuitous. Theoretical
physics took me closer to the ultimate truths of the physical
world, while my pursuit of experimental psychology was a
first step toward uncovering truth in the inner world of consciousness. Moreover, the deeper I went in these two fields,
the closer the truths of the inner and outer worlds became.

The bridge that linked them was light.

Both relativity and quantum physics, the two great paradigm shifts of modern physics, started from anomalies in
the behavior of light. Both led to radical new understand-
ings of the nature of light. Light, it seemed, occupied a very
special place in the cosmos; it was in some ways more fundamental than space, time, or matter.

Of these two paradigm shifts, the theory of relativity fascinated me the most. In high school I had pondered its implications for the nature of space and time. At university, it was my favorite part of the physics curriculum. And more recently I have realized that relativity points in exactly the same direction as Kant's arguments.

The theory of relativity sprang from the curious character of the speed of light. According to classical physics, measurements of the speed of light should vary according to the motion of the observer. Such variations happen all the time in everyday life. For example, if you are cycling along a road at 20 m.p.h., and a car traveling at 30 m.p.h. passes you, then, relative to you, the car is traveling at 10 m.p.h. If you were to pedal a little faster until you were also moving at 30 m.p.h., the car's speed relative to you would be zero, and you would be able to have a conversation with the driver.

Light moves millions of times faster than a bike, so you wouldn't expect to notice any significant differences in its speed relative to you. Nevertheless, you would expect the same principle to apply. The faster you traveled, the slower would be the speed of light relative to you. But when physicists tried to detect these changes, they obtained puzzling results. Whether traveling toward the light or away from it, the relative speed of light was always the same.

Perplexed by these findings, two American scientists, Albert Michelson and Edward Morley, designed an experiment that could detect variations in the speed of light to an accuracy of two miles per second, which was about a hundred times more accurate than the expected variation. Yet they still came up with exactly the same result. The observed speed of light never varied.

For the existing scientific paradigm, this was a major anomaly. Why did light not obey the same laws as everything else? It just didn't make sense.

Einstein's Paradigm Shift

Enter the young Albert Einstein. Having failed his college entrance examinations in electrical engineering, and having been turned down for various teaching posts in mathematics and physics, he had finally gained employment as an "assistant, third class," in the Swiss patent office. During his spare time he pondered various mathematical and physical problems, including the inexplicable results of the Michelson-Morley experiment.

In 1905, at the age of 26, while virtually unknown to the scientific community, he published two seminal papers, one on the quantum nature of light, which we will examine shortly, and "The Electrodynamics of Moving Bodies," in which he proposed a radical resolution to the problem of the speed of light, laying the foundations for his special theory of relativity.[1]

The basic premise of relativity was not new. Two hundred and fifty years earlier, Galileo had realized that if you

---

[1]Einstein called it the *special* theory of relativity to distinguish it from the general theory of relativity, which deals with gravity and the curvature of space and time.

were in a closed room with no windows, there would be no way of telling whether the room was at rest or moving with a steady velocity; any experiment you were to perform in a moving room would have the same results as one performed in a stationary room.

Imagine, for example, you are flying in a plane and you drop a tennis ball. The ball will fall vertically (from your perspective) to the floor and bounce up again towards your hand. It does not slam into the rear of the plane at 500 miles per hour. Relative to you, the ball behaves in the same way as it would if you were standing on the ground. You cannot tell from the ball's motion alone whether the plane is moving or at rest.

Galileo's theory—now known as *classical relativity*—states that the laws of physics are the same in all uniformly moving frames of reference. The phrase "uniformly moving" is important. It means moving at a steady speed in a steady direction. If the plane were accelerating or turning, you could tell that you were moving. The ball would roll across the floor, and you might feel changes in the pressure of the seat against your body.

Classical relativity referred to the motion of physical objects; it said nothing about light. Einstein took classical relativity and brought it up to date. He proposed that the principle of relativity should be valid for *all* the laws of physics, including those governing light. These, too, should be the same in all uniformly moving frames.

In 1864, James Clark Maxwell had proposed that light consisted of electromagnetic waves, with their own equations of motion. These equations specified a precise value for the speed of light of 186,282 miles per second (about 670,000,000 miles per hour). If, as Einstein argued, these equations are the same in all uniformly moving frames of reference, then the speed of light must be the same in all such frames.

In other words, however fast you are moving you will *always* measure the speed of light to be 186,282 miles per second—just as Michelson and Morley had found. Even if you were to travel at 186,281 miles per second, light would not pass by a mere 1 mile per second faster; it would still zoom by at 186,282 miles per second. You would not have caught up with light by even the tiniest amount.[2]

This goes totally against common sense. But in this instance it is common sense that is wrong. Our mental models of reality have been derived from a lifetime's experience of a world where velocities are far below the speed of light. At speeds close to that of light, reality is very different.

---

[2]This is the speed of light in a vacuum. Light is slowed when passing through a medium such as glass or water, which is why the bottom of a swimming pool appears closer than it is, and why a prism or lens can bend light.

Also remember that, to a physicist, light is more than just the light we see with our eyes. It is the whole spectrum of electromagnetic radiation, of which the visible spectrum is just one tiny range of frequencies.

The Relativity of Space and Time

That the speed of light is the same for all observers—however fast they are moving—is strange enough, but even stranger things are in store for our notions of space and time.

Einstein's equations of motion predict that moving clocks will run slower than clocks that are at rest. At the speeds we usually encounter, the difference is negligible. But as we approach the speed of light the effect becomes quite noticeable. If you were to travel past me at 87 percent of the speed of light, I would observe your clocks running at half the speed of mine. This slowing applies not just to man-made clocks, but to all physical processes, to all chemical processes, and to all biological processes. Your whole world appears to run slower than mine. Time itself is running slower.

Weird as this may seem, experiments have shown that this slowing of time actually does happen. Very sensitive atomic clocks have been flown around the world, and they have been found to run slow by exactly the predicted amount. The change is very small—a factor of about one-trillionth—but it is there.

Nor is it just time that changes; space is also affected. As an observer approaches the speed of light, measurements of length (that is, measurements of space in the direction of motion) get shorter, and in exactly the same proportion as time slows. If you were passing by me at 87 percent of the speed of light, lengths in your universe would have shrunk to half of mine.

Again this seems to defy common sense; space, like time, seems fundamental and fixed, not something that changes according to your speed. Nevertheless, experiments with subatomic particles traveling at speeds close to that of light have verified the effect. The faster you go, the more compressed space becomes.

Henceforth, space by itself, and time by itself, are doomed to fade away into mere shadows, and only a kind of union of the two will preserve an independent reality.

Herman Minkowski

The Realm of Light

For an observer actually traveling at the speed of light, the equations of special relativity predict that time would come to a complete standstill, and length would shrink to nothing. Physicists usually avoid considering this strange state of affairs by saying nothing can ever attain the speed of light, so we don't have to worry about any bizarre things that might occur at that speed.

When physicists say nothing can ever attain the speed of light, they are referring to objects with mass. Einstein showed that not only do space and time change as speed increases, so does mass. In the case of mass, however, the change is an increase rather than a decrease; the faster something moves, the greater its mass becomes. If an object were ever to reach the speed of light its mass would become infinite. However, to move an infinite mass would take an infinite amount of energy, more energy than there is in the entire universe. Thus, it is argued, nothing can ever attain the speed of light.

Nothing, that is, except light. Light travels at the speed of light. And it can do so because it is not a material object; its mass is always precisely zero.

Since light travels at the speed of light, let's imagine a disembodied observer (pure mind with no mass) traveling at the speed of light. Einstein's equations would predict that, from light's own point of view, it travels no distance and takes zero time to do so.

This points toward something very strange indeed about light. Whatever light is, it seems to exist in a realm where there is no before and no after. There is only *now*.

The Quantum of Light

More hints as to what light is—and what light is not—are found in the other great paradigm shift of modern physics, quantum theory. As with relativity, the anomaly that sparked this shift concerned light.

When you raise the temperature of a metal rod, it begins to glow a dull red. As it gets hotter, the color brightens and changes from red to orange, then to white, and finally takes on a bluish tinge. Why should this be? According to classical physics, all glowing bodies should radiate the same color, whatever their temperature.

In 1900, the German physicist Max Planck realized he could account for these changes in color if energy was not radiated in a continuous smooth flow, as had previously been supposed, but came in discrete packets, or *quanta* (from the Latin word *quantum*, meaning "amount"). He proposed that any energy exchange, whether it be an electron in an atom changing its orbit, or the warming of your skin from sunlight, consisted of a number of whole quanta. The energy change could involve 1, 2, 5, or 117 quanta; but not half a quantum or 3.6 quanta. When Planck applied this constraint to the light radiated from a glowing object, he found it led precisely to the changes in color that are observed.

Five years later, in the same year as he published his theory of special relativity, Einstein came to a similar conclusion. He was exploring the newly discovered photoelectric effect, in which light shining on a metal can trigger the release of electrons. The only way he could explain the rate at which electrons appeared was to assume that light was transmitted as a stream of particles, or *photons*. Each of these photons of light was equivalent to one of Planck's quanta, or packets of energy.

Light as Action

A quantum may be the smallest packet of energy that can be transmitted, but the energy contained in a quantum varies considerably. A gamma-ray photon, for example, packs billions of times more energy than an infrared photon. This is why gamma rays, X-rays, and even ultraviolet light to some extent, can be so dangerous. When these photons hit your body, the energy released can blow apart the molecules in a cell. On the other hand, when an infrared photon is absorbed by the body, the energy released is far less; all it does is vibrate the molecules, warming you a little.

Although the amount of energy in a photon varies enormously, there is one aspect of the quantum that is fixed. Each and every quantum has a constant amount of *action*.

Mathematicians define action as an object's momentum multiplied by the distance it travels, or the object's energy multiplied by the time it is traveling—the two are equivalent. The amount of action in a ball thrown across a football field, for example, would be greater than the same ball thrown half the distance. Double the ball's mass, and you double the action. Or imagine yourself running with a constant rate of energy output. If you run for twice as long, there will be twice the action—which makes sense intuitively.

The amount of action in a quantum is exceedingly small, about 0.00000000000000000000000000662618 erg.secs (or 6.62618×10^{-27} erg.secs in mathematical shorthand)—but it is always exactly the same amount.[3]

---

[3]The erg is a unit of energy. To lift a weight of one pound a height of one foot requires about 13.5 million ergs; so it is an extremely small unit of energy. If you took one second to lift the one pound weight, the total action involved would be 13.5 million erg.secs. That's about two billion trillion trillion quanta of action, showing just how tiny a quantum is.

This is called Planck's constant (after its discoverer). It is the second universal constant to emerge from modern physics. Like the first—the speed of light—it is a constant of light. Light always comes in identical units of action.

All matter is just a mass of stable light.
Sri Aurobindo

Like relativity, quantum theory also points to light as being beyond space and time. We may think of a photon as being emitted from some point in space and traveling to another point where it is absorbed. But quantum theory says that we know nothing of what happens on the way. The photon cannot even be said to exist in between the two points. All we can say is that there is a point of emission, a corresponding point of absorption, and the transfer of a unit action between the two.

Unknowable Light

Kant argued that the *noumenon*—the "thing-in-itself," the physical reality that is apprehended by the senses and interpreted by the mind, but never experienced directly—transcended space and time.

A hundred and twenty years later, we find Einstein lending support to Kant. Time and space are not absolutes. They are but two different appearances of a deeper reality, the *spacetime continuum*—something beyond both space and time, but with the potential to manifest as both space and time. But the spacetime continuum itself, like Kant's noumenon, is never directly known.

If we think we can picture what is going on in the quantum domain, that is one indication that we've got it wrong.

Werner Heisenberg

Light, too, has unknowable qualities. We never see light itself. The light that strikes the eye is known only through the energy it releases. This energy is translated into a visual image in the mind. Although the image appears to be composed of light, the light we see is a quality appearing in consciousness. What light actually is, we never know directly.

Light seems to lie beyond reason and commonsense understanding, a finding that again parallels Kant's conjectures. Reason, he said, was not an intrinsic quality of the noumenon, but was, like space and time, part of the way the mind made sense of things. If so, it should not be surprising that our minds find it so hard to comprehend the nature of light. It may be that we will never be able to make sense of it. With light, we may have reached the threshold of knowability.

6

The Light of Consciousness

The one "I am" at the heart of all creation,
Thou art the light of life.

My studies in experimental psychology had taught me much about neurophysiology, memory, behavior, and perception. Yet, despite all I was learning about brain function, I was no closer to understanding the nature of consciousness itself. The East, however, appeared to have a lot to say about the subject, and so did many mystics from around the world. For thousands of years such seekers had focused on the inner realm of the mind, exploring its subtler aspects through direct personal experience.

Believing that such approaches might offer insights unavailable to Western science, I began delving into ancient texts such as *The Upanishads, The Tibetan Book of the Great Liberation, The Cloud of Unknowing,* and contemporary writers such as Alan Watts, Aldous Huxley, Carl Jung, and Christopher Isherwood.

I was fascinated to find that as in modern physics, light was a recurrent theme. Consciousness itself was often characterized in terms of light. *The Tibetan Book of the Great Liberation* described "the self-originated Clear Light, eternally unborn . . . shining forth within one's own mind." St. John referred to "the true light, which lighteth every man that cometh into the world."

With all your science can you tell how it is, and whence it is, that light comes into the soul?

Henry David Thoreau

Those who have awakened to the truth about reality—whom we often call illumined, or enlightened—frequently describe their experiences in terms of light. The Sufi Abu 'l-Hosian al-Nuri experienced a light "gleaming in the Unseen. . . . I gazed at it continually, until the time came when I had wholly become that light."

The tenth-century Christian mystic St. Symeon saw

a light infinite and incomprehensible . . . one single light . . . simple, non-composite, timeless, eternal . . . the source of life.

The more I explored this inner light, the more I saw close parallels with the light of physics. Physical light has no mass, and is not part of the material world. The same is true of consciousness; it is immaterial. Physical light seems to be fundamental to the universe. The light of consciousness is likewise fundamental; without it there would be no experience.

I began to wonder whether there was some deeper significance in these similarities. Were they pointing to a more fundamental connection between the light of the physical world and the light of consciousness? Do physical reality and the reality of the mind share the same common ground—a ground whose essence is light?

Meditation

It was obvious that I would not answer such questions through mere argument and reason. As both Eastern philosophy and mystical writings make very clear, knowledge of subtler levels of consciousness comes not from reading, or from studying the experiences of others, but from one's own direct experience. So I began to look into meditation and other spiritual practices.

It happened that several Buddhist teachers and Tibetan lamas were teaching in Cambridge, including Trungpa Rinpoche, who had recently escaped from the Chinese invasion. At that stage in my exploration, Buddhism appealed to me because it was the most non-religious of the Eastern philosophies. It was as much a psychology and a philosophy as a religion. It made a point of not discussing God; its focus was removing the causes of suffering in oneself. So I started attending classes in Buddhist meditation, listening to various teachers, and reading some of the great Buddhist texts.

Several months later, the direction of my inner exploration took an unexpected turn. Hunting through the esoteric section of my local library for works on consciousness, I noticed a book entitled *The Science of Being and Art of Living* by Maharishi Mahesh Yogi, the Indian teacher who had recently made the headlines when the Beatles renounced their use of drugs in favor of his technique of Transcendental Meditation.[1] I added the book to my pile and took it back to my study, where it sat unopened on my desk for two weeks. Finally, not suspecting how much my life was about to change, I took a look. Within minutes I was riveted. Maharishi was saying the exact opposite of nearly everything I'd heard or read about meditation, yet he seemed to make perfect sense.

Most of the works I had read on meditation described how much effort was required to still the restless mind and achieve a state of deep inner peace and fulfillment. Maharishi viewed the process in a different way. The least bit of trying, even a desire for the mind to settle down, would, he observed, be counterproductive. Any effort would promote mental activity rather than lessening it.

---

[1] The publicity the Beatles gave to Transcendental Meditation (TM) may have been one of their greatest legacies. As I travel the world, I am repeatedly astonished by the number of people whose first spiritual practice was TM, back in the sixties or seventies.

He suggested that the mind was restless because it was seeking something—namely, greater satisfaction and fulfillment. But it was looking for it in the wrong direction, in the world of thinking and sensory experience. All that was needed, he said, was to turn the attention inward and then, applying his technique, allow the mind to settle down just a little. Being in a slightly quieter state, the mind would get a taste of the fulfillment it had been seeking. By repeating the practice, it would be spontaneously drawn on to yet quieter and more fulfilling levels of its own accord.

*In the final analysis
the hope of every person is simply
peace of mind.*

The Dalai Lama

Maharishi's ideas appealed to my scientific mind. They were simple and elegant—almost like a mathematical derivation. But the skeptic in me was not going to accept anything on faith. The only way to know how well his technique worked was to try it.

The nearest teacher I could find was in London, so I traveled down from Cambridge each day for a week to take some instruction. It took some time to get the practice right, but once I did, I realized Maharishi was correct. The less I tried, the quieter my mind became.

Journey to India

The following summer, I traveled to Lago di Braies, a lake high up in the Italian Alps, for a meditation retreat with Maharishi. I was instantly charmed. With his deep, warm, brown eyes, long flowing black hair and beard, dressed only in a single sheet of white cotton artfully wrapped around his small body and a simple pair of sandals, he looked the classic Indian guru. Bubbling over with joy, he never tired of talking to us novices about finer levels of being and higher states of consciousness. This was not book knowledge, but wisdom that was coming from someone who clearly had direct personal experience of these states. I knew then that I wanted to study further with him.

As soon as I completed my undergraduate degree, I earned some money driving a truck, then set off overland for India. My destination was Rishikesh, an Indian holy town at the foot of the Himalayas, about 150 miles north of Delhi.

The plains of Northern India do not gradually rise up into mountains as do the Alps; the landscape looks more like the Rocky Mountains in Colorado. One moment it is flat, the next moment there are mountains. Rishikesh nestled right where plain became mountain, at the very point where the Ganges River tumbled out of its deep Himalayan gorge.

On one side of the river was Rishikesh the bustling market town, its crowded streets a jumble of market stalls, honking cars, bicycle rickshaws, and bony cows. On the other side was Rishikesh the holy town. The atmosphere there was very different. There were no cars; the one bridge across the river—strung high above the mouth of the gorge—was deliberately built too narrow for cars. Along this side of the river, and sprinkled up the jungle hillsides, were all manner of ashrams. Some were austere walled quadrangles lined with simple meditation cells; others were graced with lush gardens, fountains, and brightly colored statues of Indian deities. Some were centers for hatha yoga, some for meditation; others were devoted to a particular spiritual teacher or philosophy.

Two miles downriver from the bridge was Maharishi's ashram, the last habitation before the winding track disappeared into the jungle. Perched on a cliff top a hundred feet above the swirling Ganges were a half dozen bungalows, a meeting hall, dining room, showers, and other facilities providing basic Western comforts.

There a hundred of us, of all ages, from many countries, had gathered for a teacher training course. Many were like myself, recent graduates looking for a deeper intellectual understanding of Maharishi's teachings as well as a deeper experience of meditation. Our group included Ph.D.'s in philosophy, physicians, and long-term students of theology.

Over the following months we listened to Maharishi expound his philosophy. We asked question after question, virtually interrogating him at times. We wanted to tease out everything, from the finer distinctions of higher states of consciousness and subtle influences of meditation, to the exact meaning of various esoteric concepts. Maharishi, in his willingness to share his knowledge, never tired. Often, when the day's program was complete, a few of us would gather in his small sitting room, where we stayed late into the night absorbing yet more of his wisdom.

Pure Consciousness

As well as furthering our understanding of meditation, Maharishi wanted us to have clear experiences of the states of consciousness he was describing. That could only come from prolonged periods of deep meditation. At first we meditated for three or four hours a day, but as the course progressed, our practice times increased. Six weeks into our three-month stay, we were spending most of the day in meditation—and much of the night as well.

> Return to the root is called Quietness;
> Quietness is called submission to Fate;
> What has submitted to Fate has become
> part of the always-so;
> To know the always-so is to be Illumined.
>
> Tao Te Ching

During these long meditations, my habitual mental chatter began to fade away. Thoughts about what was going on outside, what time it was, how the meditation was progressing, or what I wanted to say or do later, occupied less and less of my attention. Random memories of the past no longer flitted through my mind. My feelings settled down, and my breath grew so gentle as to virtually disappear. Mental activity became fainter and fainter, until finally my thinking mind fell completely silent. In Maharishi's terminology, I had *transcended* (literally "gone beyond") thinking.

Indian teachings call this state *samadhi,* meaning "still mind." They identify it as a state of consciousness fundamentally different from the three major states we normally experience: waking, dreaming, and deep sleep. In waking consciousness we are aware of the world perceived by the senses. In dreaming we are aware of worlds conjured by the imagination. In deep sleep there is no awareness, neither of outer world nor inner world. In samadhi there is awareness—one is wide awake—but there is no object of awareness. It is pure consciousness, consciousness before it takes on the various forms and qualities of a particular experience.

Yoga is the cessation of the modifications of mindstuff.

Patanjali

In the analogy of a film projector, this fourth state of consciousness corresponds to a projector running without any film, so that only white light falls on the screen. Likewise, in samadhi there is the light of pure consciousness, but nothing else. It is the faculty of consciousness without any content.

The Isha Upanishad, an ancient Indian text, says of this fourth state:

> It is not outer awareness,
> It is not inner awareness,
> Nor is it a suspension of awareness.
> It is not knowing,
> It is not unknowing,
> Nor is it knowingness itself.
> It can neither be seen nor understood,
> It cannot be given boundaries.
> It is ineffable and beyond thought.
> It is indefinable.
> It is known only through becoming it.

Similar descriptions can be found in almost every culture of the world. Using remarkably similar terms, the fifth-century Christian mystic Dionysius described it this way:

> It is not soul, or mind . . .
> It is not order or greatness or littleness . . .
> It is not immovable nor in motion nor at rest . . .
> Nor does it belong to the category of non-existence,
> or to that of existence . . .
> Nor can any affirmation or negation apply to it.

The Buddhist scholar D.T. Suzuki referred to it as a "state of Absolute Emptiness":

> There is no time, no space, no becoming, no thingness.
> Pure experience is the mind seeing itself as reflected in
> itself. . . . This is possible only when the mind is *sunyata*

[emptiness] itself; that is, when the mind is devoid of all its possible contents except itself.

The Essence of Self

When the mind is devoid of all content, we not only find absolute serenity and peace, we also discover the true nature of the self.

Usually we derive our sense of self from the various things that distinguish us as individuals—our bodies and their appearance, our history, our nationality, the roles we play, our work, our social and financial status, what we own, what others think of us. We also derive an identity from the thoughts and feelings we have, from our beliefs and values, from our creative and intellectual abilities, from our character and personality. These, and many other aspects of our lives, contribute to our sense of who we are.

However, such an identity is forever at the mercy of events, forever vulnerable, and forever in need of protection and support. If anything on which our identity depends changes, or threatens to change, our very sense of self is threatened. If someone criticizes us, for example, we may feel far more upset than the criticism warrants, responding in ways that have more to do with defending or reinforcing our damaged self-image than with addressing the criticism itself.

In addition to deriving an identity from how we experience ourselves in the world, we also derive a sense of self from the very fact that we are experiencing. If there is experience, then there must, we assume, be an experiencer; there must be an "I" who is doing the experiencing. Whatever is going on in my mind, there is this sense that I am the subject of it all.

But what exactly is this sense of "I-ness"? I use the word "I" hundreds of times a day without hesitation. I say that I am thinking or seeing something, that I have a feeling or desire, that I know or remember something. It is the most familiar, most intimate, most obvious aspect of myself. I know exactly what I mean by "I"—until I try to describe it or define it. Then I run into trouble.

Looking for the self is rather like being in a dark room with a flashlight, shining it around trying to find the source of the light. All one would find are the various objects in the room that the light falls upon. It is the same when I try to look for the subject of all experience. All I find are the various ideas, images, and feelings that the attention falls upon. But these are all objects of experience; they cannot therefore be the subject of the experience.

What is this "I"?. . . You will, on close intro-spection, find that what you really mean by "I" is the ground-stuff upon which [experi-ences and memories] are collected.

Erwin Schrödinger

Although the self may never be known as an object of experience, it can be known in another, more intimate and immediate way. When the mind is silent, and the thoughts, feelings, perceptions, and memories with which we habitually identify have fallen away, then what remains is the essence of self, the pure subject without an object. What we then find is not a sense of "I am this" or "I am that," but just "I am."[2]

In this state, you know the essence of self, and you know that essence to be pure consciousness. You know this to be what you really are. You are not a being who is conscious. You are consciousness. Period.

I AM You, that part of you who is and knows . . .
that part of you who says I AM and is I AM. . . .
I AM the innermost part of you that sits within,
and calmly waits and watches, knowing neither
time nor space. . . .
It was I Who directed all your ways, Who inspired
all your thoughts and acts. . . .
I have been within always, deep within your
heart.

The Impersonal Life

This core identity has none of the uniqueness of the individual self. Being beyond all attributes and identifying

---

[2]Even to say "I am" can be misleading; the word "I" already has so many associations with an individual self. It might be more accurate to say there is *amness,* or pure being.

characteristics, *your* sense of I-ness is indistinguishable from *mine*. The light of consciousness shining in you, which you label as "I," is the same light that I label as "I." In this we are identical.

I am the light. And so are you.

Beyond Time and Space

This essential self is eternal; it never changes. It is pure consciousness, and pure consciousness is timeless.

Our normal experience of the passing of time is derived from change—the cycle of day and night, the beating of the heart, the passing of thoughts. In deep meditation, when all awareness of things has ceased and the mind is completely still, there is no experience of change, and nothing by which to mark the passing of time. I may know I have been sitting in absolute stillness, but as to how long I have been there, I may have no idea. It could have been a minute, or it could have been an hour. Time as we know it disappears. There is simply now.

Time and space are but physiological colors which the eye makes, but the Soul is light.
Ralph Waldo Emerson

Not only is the essential self beyond time, it also is beyond space. If we are asked to locate our own consciousness, most people sense it to be somewhere in the head. Right now this book probably appears a couple of feet in front of you. You may be aware of walls around you; the ground some feet below you; and your arms, torso, legs, and feet are also out there, a short distance from the point of your perceiving self.

The feeling that our consciousness is located somewhere in the head seems to make sense. Our brains are in our heads, and the brain is somehow associated with conscious experience. We would find it strange if, for example, the brain was in the head, but our consciousness was in our knees.

However, all is not as it seems. The apparent location of consciousness does not actually have anything to do with the placement of the brain. It depends on the placement of the sense organs.

Our primary senses, our eyes and ears, happen to be situated on the head. Thus the central point of our perception, the point from which we seem to be experiencing the world, is somewhere behind the eyes and between the ears—somewhere, that is, in the middle of the head. The fact that our brains are also in our heads is just a coincidence, as the following simple thought experiment bears out.

Imagine that your eyes and ears were transplanted to your knees, so that you now observe the world from this new vantage point. Where would you now experience your self to be—in your head or down by your knees? Your brain may still be in your head, but your head is no longer the central point of your perception. You would now be looking out onto the world from a different point, and you might well imagine your consciousness to be in your knees.[3]

Our perceiving self is nowhere to be found within the world-picture, because it itself is the world-picture.

Erwin Schrödinger

In short, the impression that your consciousness exists at a particular place in the world is an illusion. Everything we experience is a construct within consciousness. Our sense of being a unique self is merely another construct of the mind. Quite naturally, we place this image of the self at the center of our perceived world, giving us the sense of being *in* the world. But the truth is just the opposite: It is all within us.

You have no location in space. Space is in you.

The Universal Light

Here again we see close parallels between the light of consciousness and the light of physics. When we considered physical

---

[3]This sheds new light on so-called "out-of-body" experiences in which we find ourselves experiencing the world from a different vantage point—looking down on ourselves from the ceiling, for example. The central point of perception is no longer in the same vicinity as the body. We think we have left the body, but the truth is we were never *in* our bodies in the first place.

light from its own frame of reference, we found that distance and time disappeared. The realm of light seems to be somehow beyond space and time. Likewise, when we consider the nature of pure consciousness, space and time disappear. In both cases there is only the ever-present moment.

In physics, light turns out to be absolute. Space, time, mass, and energy are not as fixed as we once thought they were. The new absolutes are those of light—the speed of light in a vacuum and the quantum of action of a photon. Similarly, in the realm of mind, the faculty of consciousness is absolute. It is the common ground of all experience— including that of space and time. Consciousness itself, like the light in a projector, is unchanging, eternal.

Every photon of light is an identical quantum of action. The same is true of consciousness. The light of consciousness shining in me is the same light that shines in you—and in every sentient being.

Those wise ones who see that the consciousness within themselves is the same consciousness within all conscious beings, attain eternal peace.

Katha Upanishad

These parallels suggest there may be some deeper relationship between the light of physics and consciousness. Could they share a common ground—a common ground that manifests in the physical realm as light, and in the realm of mind as the light of consciousness that shines in every being?

God's first command in Genesis was "Let there be light," and from this light the whole of creation was born. It might, however, be more accurate to say, "the whole of creation *is* born," for light underlies everything that happens. This is true in the physical world, where every interaction involves the exchange of photons. And it is true in the subjective realm, where the light of consciousness is the common ground of every experience.

God is the light of Heaven and of the Earth.

Quran

I am not suggesting that light *is* God, but that light may be the first manifestation of the underlying ground of all existence, the subtlest level of creation, the closest we can come to that which lies beyond all form. In the realm of conscious experience, the pure self—the inner light that lies behind the countless forms arising in the mind—is where we touch the divine. This explains why many of those who have explored deep within and discovered their true nature have made one of the most contentious and confusing of all mystical claims—the assertion that "I am God."

7

Consciousness as God

The soul is in itself a most lovely and
perfect image of God.

St. John of the Cross

To many, the statement "I am God" rings of blasphemy. God, according to conventional religion, is the supreme deity, the almighty eternal omniscient creator. How can any lowly human being claim that he or she is God?

When the fourteenth-century Christian priest and mystic Meister Eckhart preached that "God and I are One," he was brought before Pope John XXII and forced to recant all such teachings. Others suffered a worse fate. The tenth-century Islamic mystic Al-Hallaj was crucified for using language that claimed an identity with God.

Yet when mystics say "I am God," or words to that effect, they are not talking of an individual person. Their inner explorations have revealed the true nature of the self, and it is this that they identify with God. They are claiming that the essence of self, the sense of "I am" without any personal attributes, is God.

The contemporary scholar and mystic Thomas Merton put it very clearly:

> If I penetrate to the depths of my own existence and my own present reality, the indefinable *am* that is myself in its deepest roots, then through this deep center I pass into the infinite *I am* which is the very Name of the Almighty.

"I am" is also one of the Hebrew names of God, Yahweh. Derived from the Hebrew YHWH, the unspeakable name of God, it is often translated as "I AM THAT I AM."

> *I am the infinite deep*
> *In whom all the worlds appear to rise.*
> *Beyond all form, forever still.*
> *So am I.*
>
> Ashtavakra Gita

Similar claims appear in Eastern traditions. The great Indian sage Sri Ramana Maharshi said:

> "I am" is the name of God . . . God is none other than the Self.

In the twelfth century, Ibn-al-Arabi, one of the most revered Sufi mystics, wrote:

> If thou knowest thine own self, thou knowest God.

Shankara, the eighth-century Indian saint, whose insights revitalized Hindu teachings, said of his own enlightenment:

> I am Brahman. . . . I dwell within all beings as the soul,
> the pure consciousness, the ground of all phenomena. . . .
> In the days of my ignorance, I used to think of these as
> being separate from myself. Now I know that I am All.

This sheds new light on the Biblical injunction "Be still, and know that I am God." It does not mean "Stop fidgeting around and recognize that the person who is speaking to you is the almighty God of all creation." It makes much more sense as an encouragement to still the mind and know, not as an intellectual understanding but as a direct realization, that the "I am" that is your essential self, the pure consciousness that lies behind all experience, is the supreme Being, the source of all.

This concept of God is not of a separate being, beyond us in some other realm, overlooking human affairs and loving or judging us according to our deeds. God appears in each and every one of us as the most intimate and undeniable aspect of ourselves, the consciousness shining in every mind.

I Am the Truth

When I discovered that many sages and mystics described their experience of pure consciousness as a personal knowing of the divine, several traditional descriptions of God became clearer to me.

The faculty of consciousness is, as we have seen, the only absolute, unquestionable truth. Whatever is taking place in my mind, whatever I may be thinking, believing, feeling, or sensing, the one thing I cannot doubt is consciousness. God, likewise, is often said to be the one absolute truth.

God is universal. So is the faculty of consciousness. It is a primary quality of the cosmos, an intimate aspect of all existence.

Like God, consciousness is omnipresent. As the familiar saying goes, "Wherever you go, there you are." Whatever your experience, whatever forms arise in the mind, the sense of "amness" is always present. It has always been present; it will always be present. It never changes. It is eternal and everlasting.

---

*When I say "I am," I do not mean a
separate entity with a body as its nucleus.
I mean the totality of being, the ocean of
consciousness, the entire universe of all
that is and knows.*

Sri Nisargadatta Maharaj

God is often said to be the creator and the source of all creation. So is consciousness. Our entire personal world—everything we see, hear, taste, smell, and touch; every thought, feeling, fantasy, intimation, hope, and fear—is a form that consciousness has taken on. Consciousness is the source and creator of everything we know.

Awareness of our own essential nature likewise has qualities generally associated with the divine. When the mind is silent and no longer troubled by concerns for the past or future, we connect with the pure self beyond name and form. In this experience of pure Being we find a steady, unshakable peace that is not dependent on what we have or do in life. We find the fulfillment we have always been seeking—the peace of God that passes, or lies beyond, all understanding.

This pure Mind, the source of everything,
Shines forever and on all with the brilliance
of its own perfection.
But the people of the world do not awake to it,
Regarding only that which sees, hears, feels
and knows as mind.
Blinded by their own sight, hearing, feeling
and knowing,
They do not perceive the spectral brilliance
of the source of all substance.

Huang Po

The Materialist Mindset

Not only do traditional descriptions of God make new sense when God is identified with the essence of consciousness, so do many spiritual practices.

In earlier chapters, we considered our construction of reality in terms of our sensory perception—the sounds, colors, and sensations we experience. The way in which we produce this picture of the world is more or less hard-wired into the brain.[1] How we interpret this picture, however, varies considerably. You and I may assess a person's actions in very different ways. We may read very different meanings into a news story, or see a situation at work in different lights. These varying interpretations stem from the beliefs, assumptions and expectations we bring to the situation—what psychologists call our *mind sets*.

In much the same way as our various scientific paradigms are founded on a yet more fundamental belief, or metaparadigm, the various assumptions that determine the meaning we give to our experience are based on a more fundamental mindset. We believe that inner peace and fulfillment come from what we have or do in the external world.

Tragically, this way of thinking actually prevents our finding true peace of mind. We can become so busy worrying about whether or not we may be at peace in the future, or so busy being angry or resentful about what has stood in

---

[1]There are exceptions: Some drugs can modify brain chemistry and so change the way sensory data is processed, leading to an image of reality that is different from normal—colors may shift, objects may seem less solid, space and time may change. Similar effects can happen in extremes of fatigue, sickness, stress, or in certain spiritual practices. Generally, however, if the brain is functioning normally, we all construct similar pictures of reality.

the way of peace in the past, we never have the chance to be at peace in the present.

Don't worry, be happy.

Meher Baba

The general effect of this material mindset is to put our inner state of mind at the mercy of the external world. In this respect, too, it is similar to the materialist metaparadigm of contemporary science. In both instances, consciousness is assumed to be dependent upon the material world. The current scientific worldview believes that consciousness emerges from the world of space, time, and matter. This materialist mindset tells us that our state of mind depends on events in the world of space, time, and matter. Moreover, as with the scientific metaparadigm, the mindset that runs our lives is seldom questioned.

Spirituality 101

We do not have to perceive the world through this mindset. If we perceive life from the perspective that all we know is a construct of consciousness, everything changes.

With this shift, whether or not we are at peace is no longer determined by what we have or do in the material world. We have created our perception of the world. We have given it all the meaning and value it has for us. And we are free to see it differently.

Nothing has to be achieved in order to be at peace. All we have to do is stop doing—stop wanting things to be different, stop worrying, stop getting upset when things don't go as we would wish, or when people don't behave as we think they should. When we stop doing all the things that obscure the peace that is there at our core, we find that what we have been seeking all along is there, waiting silently for us.

People are disturbed not by things, but the view they take of them.

Epictetus

This, to me, is Spirituality 101. It is a universal principle, independent of time, culture, or religious belief. It is the core principle from which many spiritual practices unfold.

Forgiveness

Consider, for example, the practice of forgiveness. The conventional understanding of forgiveness is of an absolution or pardon: "I know you did wrong, but I'll overlook it this time." But the original meaning of forgiveness is very different. The ancient Greek word for forgiveness is *aphesis*, meaning "to let go." When we forgive others we let go of the judgments we may have projected onto them. We release them from all our interpretations and evaluations, all our thoughts of right or wrong, friend or foe.

Instead we see that they are human beings caught up in their own illusions about themselves and the world around them. Like us, they feel the need for security, control, recognition, approval, or stimulus. They too probably feel threatened by people and things that prevent them from finding fulfillment. And, like us, they sometimes make mistakes. Yet, behind all these errors, there is another conscious being simply looking for peace of mind.

Even those we regard as evil are seeking the same goal. It is just that for one reason or another—who knows what pain they may have endured in their childhood, or what beliefs they may have adopted—they seek their fulfillment in ways that are uncaring, and perhaps even cruel. Deep inside, however, they are all sparks of the divine light struggling to find some salvation in this world.

Forgiveness is not something we do for the other person so much as something we do for ourselves. When we let go of our judgments of others, we let go of the source of much of our anger and many of our grievances.

There is nothing more painful than walking around with bitterness in your heart.

Hugh Prather

Our bad feelings may seem justified at the time, but they don't serve us—in fact, they usually cause more damage to ourselves than they do to the other person. The freer we are of our judgments, the more at peace we can be in ourselves.

This change in perception is the essence of a change of consciousness. When I first heard of higher states of consciousness, I imagined they would bring awareness of subtler dimensions, possibly new energies, or some other aspect of reality that was beyond my everyday perception. Over the years, I have gradually realized that enlightenment is seeing the same world, but in a different light. It is not seeing different things so much as seeing things differently.

Prayer

In every moment I have a choice as to how I see a situation. I can see it through eyes caught in the materialist mindset that worries whether or not I am going to get what I think will make me happy. Alternatively, I can choose to see it through eyes free from the dictates of this thought system.

But it is not always easy to make that choice. Once I've been caught by a fearful perception, I'm seldom aware there could even be another way of seeing things. I think my reality is the only reality.

Sometimes, however, I recognize there could be another way of seeing things, but I don't know what it is. I can't make the shift on my own; I need help. But where do I go for help? Other people are as likely to be caught in this thought system as I am. The place to go for help is deep within, to that level of consciousness that lies beyond the materialistic mindset—to the God within. I have to ask God for help. I have to pray.

When I pray in this way, I am not asking for divine intervention by an external God. I am praying to the divine presence within, to my true self. Moreover, I am not praying for the world to be different than it is. I am praying for a different perception of the world. I am asking for divine intervention where it really counts—in the mindsets that govern my thinking.

No problem can be solved from the same consciousness that created it.

Albert Einstein

The results never cease to impress me. Invariably, I find my fears and judgments drop away. In their place is a sense of ease. Whoever or whatever was troubling me, I now see through more loving and compassionate eyes.

God Is Love

Love is another quality frequently ascribed to God. This love is not to be confused with what generally passes for love in our world, which, more often than not, has its origins in the same materialist mindset that runs many other areas of our lives.

We believe that if only other people would think or behave as we want them to, we would be happy. When they don't, we may find ourselves feeling upset, angry, frustrated, or some other less-than-loving emotion. Conversely, when we meet someone who we think will satisfy our deeper needs—someone, that is, who matches our image of the perfect person—our hearts are filled with warm feelings toward them. We say we love them.

Such love is conditional. We love a person for their appearance, their manner, their intellect, their body, their talents, their smell, their dress, their habits, their beliefs and values. We love someone whom we feel is special; someone who matches our expectations, someone who will satisfy our deeper needs, someone who will make our life complete.

Such love is also fragile. If the other person gains weight, develops some annoying habit, or does not care for us as we think they should, our judgments can flip from positive to negative, and the love vanish as quickly as it came.

When love and hate are both absent,
Everything becomes clear and undisguised.
Seng-ts'an, the Third Zen Patriarch

The love of which the mystics speak is a very different form of love. It is an unconditional love, a love that does not depend on another's attributes or actions. It is not based on our wants, needs, hopes, fears, or any other aspect of the materialist mindset. Unconditional love is the love that springs forth when the mind has fallen silent, and for once we are free from fear, evaluation and judgment.

Like the peace we seek, this unconditional love is always there at our core. It is not something we have to create; it is part of our inner essence. Pure consciousness—consciousness not conditioned by the needs and concerns of an individual self—is pure love. I, in my true essence, am love.

The Golden Rule

As much as we want to feel unconditional love in ourselves, we also want others to feel that love toward us. None of us want to feel criticized, rejected, ignored, or manipulated. We want to feel appreciated, honored, and cared for. This is true not only in our intimate relationships with our partners and family, but also in our relationships with those we work with, people we meet socially, and even strangers we may encounter on the street or in an airplane. In all our relationships we want to feel respected.

If love is what we all want, then love is what we should be giving each other. But that is not always easy. Too often we are so busy trying to get love for ourselves, or holding on to the love we have, we forget that other people want exactly the same. Before long we get caught in a vicious circle that denies us the very love we seek.

If we feel hurt over something someone says or does—whether they intend to hurt us, or whether it is all our own creation—our normal response is to defend by attacking in kind. Though not the wisest or most noble response, if we believe that our happiness depends on how others behave, this is how we tend to react. If the other person is trapped in the same mindset, they are likely to respond in a similar fashion and do or say something hurtful in return.

So the vicious circle is created. On the surface it may seem that a relationship is going well; both people appear friendly; there is no open hostility. But underneath a subtle game is being played. Each person, in attempting to get the other person to be more loving, is making the other feel hurt rather than loved. It is a tragic lose-lose game, which, if sustained, can ruin the best of relationships.

As easily as the circle is set up, it can be undone. The key is simple: Give love rather than withhold it. What this means in practice is that whatever we say, and however we say it, we want the other person to feel loved and cared for rather than attacked and hurt.

If you can conduct yourself in a way that is not detrimental to others or that does not impinge on their freedom, then you are behaving according to dharma.

Sai Baba

The Buddha called this "right speech": If you cannot say something in such a way that the other person feels good on hearing it, then it is better to retain noble silence. This should not be interpreted as avoidance—"I don't know how to say what I want to say without you getting upset, so I shall just keep quiet." Expressing our thoughts and feelings is valuable, but we need to do so in ways that do not trigger the vicious circle. We should retain noble silence only so long as we need to—until we've worked out how to say what we have to say in a kind and loving manner.

Spiritual teachings often refer to this principle as the *golden rule.* "Regard your neighbor's gain as your own gain, and your neighbor's loss as your loss," says Taoism. The Koran proclaims, "No one of you is a believer until he desires for his brother that which he desires for himself." While Christ said, "All things whatsoever that ye would that men should do to you, do ye even so to them."

The key is kindness, the intent to cause no harm to others. It springs from the recognition that the light of consciousness shining in us all is divine. We honor God by honoring each other, for each and every one of us is holy.

My religion is kindness.

The Dalai Lama

Unlike the God I rejected as a youth, God as the light of consciousness neither conflicts with my scientific leanings, nor does it run counter to my intuition and reason. Indeed, it points toward an ultimate convergence of science and religion.

8

The Meeting of Science and Spirit

—☀—

> God is a pure no-thing,
>   concealed in now and here:
> the less you reach for him,
>   the more he will appear.
>
> ANGELIUS SILESIUS

I returned from India with a new understanding of God, but I was not about to advocate a return to conventional religion. I wanted to translate what the world's spiritual traditions had discovered about human consciousness into terms and practices applicable to the twentieth century.

Back at Cambridge, I was faced with the question of how to integrate this new interest into my academic life. In my final undergraduate exams in theoretical physics and experimental psychology, I had been awarded a "First Class" degree (corresponding to a *summa cum laude* in America). This achievement virtually guaranteed my acceptance for Ph.D. studies. I therefore put forward a research proposal on the subject closest to my heart—meditation. I wanted to investigate the changes in brain and body that meditation induced. But the incumbent professor of psychology was not impressed. Meditation, he told me, was not an acceptable subject of study. If I wanted to study fringe phenomena, I could work on hypnosis, but not meditation.

Somewhat discouraged, I thought I might have to take a job in computer programming after all. I had, by then, completed a postgraduate degree in computer science, and had been approached by IBM about the possibility of working in their research labs in the newly emerging field of computer graphics. Who knows how my life might have developed had I taken that route—especially considering the vital role of computer graphics in today's world. However, thanks to some unanticipated events, my career took a different path.

The Stress Lab

A week after my Ph.D. proposal was declined, a friend of mine told his father about my professor's disparaging comments on meditation. His father was professor of education at Bristol University in the west of England. A few days later, he happened to mention my story to his colleague, Ivor Pleydell-Pearce, who ran the psychology department at Bristol. The next thing I knew, I had an invitation to go down to Bristol to talk with Ivor.

Ivor's research focused on stress, and he was particularly interested in meditation as an antidote to stress. Furthermore, he had an entire laboratory that was not being used, and which he could make available to me. Did I want to come and do my Ph.D. there? Needless to say, I had no difficulty accepting. Funding soon followed, and I was off.

There are only two ways to live your life: as though nothing is a miracle, or as though everything is a miracle.

Albert Einstein

The laboratory at my disposal had a sign on the door saying "Stress Lab," which amused me, because I was doing research on relaxation, its very opposite. The lab did, however, prove extremely useful. It was full of equipment for monitoring physiological processes, the very equipment I needed for my own research. As if that were not perfect enough, the lab also contained a soundproof room. There could hardly have been a less stressful place. With its door closed, there was total silence, and when I turned out the lights, total darkness—a Himalayan cave in a laboratory. I could provide the ideal environment for experimental subjects to meditate with minimal disturbance. And at the end of a long day at work, I too had the perfect place for meditation.

Bringing Spirit Down to Earth

My studies, along with those of several researchers in the U.S., revealed that Transcendental Meditation elicited physiological changes that were the exact opposite of the stress response. Virtually every indicator of stress, from heart rate and blood pressure to body chemistry and brain activity, reversed dramatically during meditation. Herbert Benson of Harvard Medical School dubbed this the "relaxation response," and almost overnight meditation became respectable. Doctors began recommending it to patients; teachers encouraged students to take it up; even business people took lessons on the quiet.

This scientific validation of meditation also had a major impact on my own life. During my second year of research I was again approached by IBM, but not about computer graphics. They had heard about the research results and asked if I would teach TM to some of their managers.

So began my corporate career. Over the next twenty years I designed and implemented programs for a variety of companies, large and small. My work expanded beyond meditation and stress management into creativity, learning, and communication. Yet my focus was always on some form of self-development. I enjoyed taking ideas and practices I had found valuable on my own inner journey and putting them into forms that were meaningful to people whose principal concerns were managing staff, meeting corporate targets, and making deals, as well as paying the mortgage and schooling the kids.

I never spoke in spiritual terms. Most of the people I was working with would have run a mile at any hint of religion or mysticism. I reasoned that if spiritual wisdom is eternal and universal, then it should be expressible in language appropriate to the current times—the language of science and reason. For spiritual development to be acceptable it must be reasonable. It must make sense within the current worldview.

Science without religion is lame; religion without science is blind.

Albert Einstein

In Bristol I also began my writing career. In my final year there, the editor of an academic journal invited me to contribute an article on consciousness. I explained that I was a scientist, not a writer. He assured me it was his job as an editor to turn whatever I wrote into good prose. Having submitted my piece, I was surprised to hear him say that my writing was quite lucid.

A few years later I discovered the reason. My training in mathematics had borne unexpected fruit. I wrote as a mathematician, constructing a logical sequence of ideas that took the mind step-by-step toward my intended conclusion.

Before leaving Bristol I began my first book, *The TM Technique*. I wanted to tackle some of the misconceptions about TM and to integrate the spiritual aspects of meditation with the scientific research on its effects. Upon its publication the BBC invited me to produce a radio series on meditation, as a result of which I wrote a second book, *Meditation*. Two years later, a friend and I produced a new translation of the Upanishads, one of the cornerstones of Indian philosophy. My growing work in the corporate world led me to write *The Brain Book* and *The Creative Manager*. Two more books, *The Global Brain* and *The White Hole in Time*, explored the relevance of inner growth to contemporary issues, particularly the information explosion and our ever-accelerating pace of development.[1]

As I continued to explore what spiritual teachings said about consciousness, I became increasingly interested in evolution—not just biological evolution, but evolution in a larger context, from the emergence of primordial matter in the early universe to the development of human culture in modern times. I saw that along with the evolution of physical form there had been a parallel evolution of consciousness. The future development of the human species, I realized, was not headed farther out into space, but inward into the hidden depths of consciousness—and ultimately to the divine.

As I mentioned in the Introduction, this interest in the evolution of consciousness led me to conclude that the current scientific metaparadigm was incomplete and that consciousness should be included as a primary aspect of reality. Pondering the nature of paradigm shifts further, I saw that just as science had evolved through a series of paradigm shifts, so too had religion. Moreover, the two sets of shifts appeared to be heading in the same direction.

---

[1]*The Global Brain* was originally published in the U.K. as *The Awakening Earth,* and later as an updated edition entitled *The Global Brain Awakens.* A revised edition of *The White Hole in Time* was recently published as *Waking Up in Time.*

Spiritual Paradigms

The earliest religions probably date back to the time when human beings became aware that they were aware, and recognized that other people were aware. It was then only a small step to suppose that other creatures were also aware. Looking into the eyes of a bear or a crow, it was not hard to imagine that "in there" was another conscious being. The same, it was assumed, applied to plants, and natural phenomena such as rivers and mountains. They too had their own souls or spirits.

The existence of such spirits explained many things to which early peoples had no easy answer: why rains fell, why volcanoes erupted, why people fell sick, why accidents happened. If a rock rolled down the mountain, injuring a member of the tribe, it was possibly because the spirit of the mountain was angry. So they might try to appease it in some way—make an offering perhaps, or pray for forgiveness.

If we had grown up in one of these traditions, we would have taken its various beliefs for reality. Its beliefs would have constituted the paradigm of our culture—not a scientific paradigm, but a paradigm nevertheless—the worldview that shaped our perception of reality. Day-to-day experiences would have been understood within that framework. Any anomalous observations—offering sacrifices to the mountain did not always prevent rocks from falling on people—would be ignored, or incorporated in some way within the prevailing worldview.

Many Gods

As cultures evolved, so did peoples' views of these spirits. Not only did each animal and plant have its own spirit; so did entire species. There was an oak deva, a bear deity, a crow god. Other natural phenomena had their own ruling spirits—the god of thunder, the spirit of the wind, the goddess of the earth. These beings did not dwell within the physical form of a particular plant or animal, but often lived up in the sky, on the tops of mountains, or in some other faraway place.

The shift from spirits that inhabited natural forms to *supernatural* ("above nature") gods and deities signified a new religious paradigm, that of *polytheism*, or "many gods." As with the spirits of earlier religions, the existence of these gods explained many things. In Greek mythology, Apollo rode across the sky carrying the sun in a chariot drawn by four flying horses. Hercules held the world aloft. Cupid made people fall in love. These gods had many human characteristics; they could be kind, ambitious, quarrelsome, jealous, angry, or wise. Some were evil, others were forces for good.

They also took an active interest in human affairs, taking care of people in need, and administering a degree of cosmic law and order. Those who behaved badly the gods would punish, either in their own lifetimes or in the afterlife— which by then had gathered its own rich mythology—while those who showed proper repentance for misdeeds would be forgiven.

One God

The next paradigm shift was the reduction of many gods to one almighty God. Around 600 B.C., in Persia, a young man named Zarathustra (said to be born of a virgin) began preaching that there was one true God. There were still various angels, archangels, and a devil, but there was only one savior—Ahura Mazda (the "Wise Lord"). Zarathustra's teachings gave rise to the religion of Zoroastrianism (Zoroaster is Greek for Zarathustra). Although it is only a minor religion today, it paved the way for the major contemporary monotheistic traditions of Judaism, Christianity, and Islam.

Think good, do good, speak the truth.

Zarathustra

In these *monotheistic* religions God was a unique, absolute, personal being—the supreme intelligence, omnipotent and omniscient. He (for God was usually cast in male form) had not only created the natural world, but continued to watch over it and take care of its peoples.

Devotional love took on an increasingly important role. Those who loved God would receive God's love in return. Love for one's fellow human beings was likewise important—although many found it difficult to practice with those who worshipped some other God.

No God

Along with the transition from polytheism to monotheism came the emergence of *atheism*, or "no God." It might seem contradictory to have a religion without a god, but several major traditions have arisen around this theme.

In India, in the sixth century B.C., a young prince named Mahavira became disenchanted with his traditional Vedic religion, which advocated the sacrifice of innocent animals, the performance of meaningless rituals, and the belief in fictitious man-made gods. Renouncing the grand lifestyle of his palace, he wandered penniless for thirteen years seeking a better way. Then one day, while absorbed in deep meditation, he experienced a unity with all creation and a liberation from worldly woes. He consequently proclaimed himself *Jina*, "the Conqueror"—the conqueror of the mind—and encouraged his followers, the *Jains*, to attain a similar liberation through righteous living, nonviolence, and harmlessness.

Shortly afterward, another Indian prince, Siddhartha Gautama, likewise left the luxury of his palace and set out to find a way to end suffering. Six years later, in deep meditation, he too attained liberation, and was called *Buddha*— "the awakened one." Buddha realized that suffering was self-created and unnecessary, and began teaching others how to wake up and find true freedom.

During the same period, two atheistic religions arose in China. Like Jina and Buddha, Lao Tsu and Confucius both taught that people could discover truth and find inner peace without believing in any deity. They, too, advocated lives of simplicity, virtue, honesty, and above all, kindness.

"Are you a God?" they asked the Buddha.
"No," he replied.
"Are you an angel, then?" "No."
"A saint?" "No."
"Then what are you?"
Replied the Buddha, "I am awake."

Huston Smith

This fourth religious paradigm lacked some of the benefits provided by a benevolent deity. There was no longer any supernatural agent to intervene in human affairs; one's destiny was now in one's own hands. But much of the rest remained. Love, kindness, and right living were important; salvation from the sufferings of the world was still possible. In a sense there was still even a devil, but now the devil was within oneself. The goal was to liberate the mind from its self-imposed limitations—from desires, attachments, delusions, and a false sense of self.

All Is God

Along with the various polytheistic, monotheistic, and atheistic religions, another recurrent spiritual theme has been *pantheism,* meaning "God is all."

Pantheistic ideas have appeared from time to time within most cultures. The sufi mystic Ibn-al-Arabi wrote

God is essentially all things. . . . The existence of all created things is His existence. Thou dost not see, in this world or the next, anything besides God.

And Meister Eckhart preached that

God is everywhere and is everywhere complete. Only God flows into all things, their very essences. . . . God is in the innermost part of each and every thing.

God sleeps in the rock,
dreams in the plant,
stirs in the animal,
and awakens in man.

                    Sufi Teaching

In Western philosophy pantheism came to prominence in the early nineteenth century in the writings of Georg Hegel, who held not only that all existence is God, but also that the whole of history is part of God's self-realization. Similar sentiments are found in the twentieth-century philosophies of Alfred North Whitehead, Pierre Teilhard de Chardin, and Sri Aurobindo.

Einstein was a pantheist. He may not have believed in any conventional notion of God, but he did believe that

a spirit is manifest in the laws of the Universe—a spirit
vastly superior to that of man, and one in the face of
which we with our modest powers must feel humble.

Pure pantheists believe that God is the essence of all
things. Others, called *panentheists* (meaning "God is *in* all"
rather than "God is all"), believe God is in all things and
also beyond them. Some pantheists believe in the reality of
the material world; others think it is illusion. Some believe
in the existence of individual souls; others do not. But they
all reject the notion of God as a separate, supreme, supernatural being, the creator of the world and judge of human
affairs.

Many people today may be pantheists without realizing
it. Having no church, no holy text, and no gurus, pantheism is not as visible as other religions, nor is it an organiza-
tion to formally join. Many of those who have rejected their
traditional monotheistic religions, yet still believe in some
deeper divinity, might find themselves sympathetic to pantheist ideas.

With pantheism, religion has almost come full circle. The
first religions held that all things had an inner spirit, but
they projected human qualities onto these spirits. The pantheist also sees spirit in everything, but a divine spirit rather
than one with human qualities and frailties.

Clearly, pantheism is not so different from the panpsychism discussed in chapter 3. Indeed, if we identify God
with the faculty of consciousness, then the view that

consciousness is in everything becomes the view that God is in everything.

Converging Paradigms

The worldviews of science and spirit have not always been as far apart as they are today. Five hundred years ago, there was little difference between them. Science, limited though it was, existed within the established worldview of the Christian church. Following Copernicus, Descartes, and Newton, Western science broke away from the doctrines of monotheistic religion, establishing its own atheistic worldview, which today is very different indeed from that of traditional religion. But the two can, and I believe eventually will, be reunited. Their meeting point is consciousness. When science sees consciousness to be a fundamental quality of reality, and religion takes God to be the light of consciousness shining within us all, the two worldviews start to converge.

Nothing is lost in this convergence. Mathematics remains the same; so do physics, biology, chemistry. The shift may throw new light on some of the paradoxes of relativity and quantum theory, but the theories themselves do not change. This inclusion is a common pattern in paradigm shifts: the new model of reality includes the old as a special case. Einstein's paradigm shift makes no difference to observers traveling at everyday speeds; as far as we are concerned Newton's laws of motion still apply. In a similar way, seeing consciousness as a fundamental quality of reality does not change our understanding of the physical world. It does, however, bring us a deeper appreciation of ourselves.

The same integrity is maintained on the spiritual side. Much of the wisdom accumulated over the ages remains unchanged; forgiveness, kindness, and love are as important as they ever were. Many of the qualities traditionally ascribed to God remain, being equally applicable to the faculty of consciousness. The difference is that spiritual teachings and scientific knowledge now share a common ground. This is another common pattern in paradigm shifts. Newton brought terrestrial and celestial mechanics under the same laws. Maxwell integrated electricity, magnetism, and light in a single set of equations. With the shift to a consciousness metaparadigm the integration goes much further. It brings the two halves of humanity's search for truth together under the same roof.

This meeting of science and spirit is crucial, not just for a more comprehensive understanding of the cosmos, but also for the future of our species. Today, more than ever, we need a worldview that validates spiritual inquiry, for it is the spiritual aridity of our current times that lies behind so many of our crises.

9

The Great Awakening

Thank God our time is now, when wrong
Comes up to face us everywhere,
Never to leave us 'til we take
The longest stride of soul man ever took.
Affairs are now soul size.
The enterprise is exploration into God.

CHRISTOPHER FRY

The more I have studied the nature of consciousness, the more I have come to appreciate the critical role that inner awakening plays in the modern world—a world which, despite all its technological prowess, seems to be getting deeper and deeper into trouble.

Most of today's problems—from personal worries to social, economic, and environmental issues—stem from human actions and decisions. These arise from human thinking, human feelings, and human values, which in turn are influenced by our belief that happiness comes from what we have and do, and by our need to bolster an ever-vulnerable sense of self. Psychological issues such as these lie at the root of our problems. The growing crises we observe around us are symptoms of a deeper inner crisis—a crisis of consciousness.

This crisis has been a long time coming. Its seeds were sown thousands of years ago when human evolution made the leap to self-awareness, and consciousness became conscious of itself.

The first appearance of self-awareness probably involved a sense of identity with one's tribe and kin, but not a strong personal self. Gradually this inner awareness evolved, becoming more focused, until today it has reached the point at which we have a clear sense of being a unique self, distinct from others and the natural environment.

If men and women have come up from the beasts, then they will likely end up with the gods.

Ken Wilber

Awareness of this individual self is not, however, the final stage of our inner evolution. Dotted through history have been those who have discovered there is much more to consciousness than most of us usually realize. This self, they tell us, is not our true identity. Moreover, it has serious shortcomings. If our awareness of self is limited to this separate, dependent, ever-vulnerable self, our thinking is distorted, and our actions are misguided, bringing much unnecessary suffering upon ourselves. To free ourselves from this handicap, we must take a further step in our inner journey and discover the true nature of consciousness.

Our Final Exam

In the past, greater awareness of the true self was deemed important for personal well-being. Today the game has changed; it is now imperative for our collective survival.

Our knowledge of the external world has been growing at an accelerating pace, bringing with it an unprecedented ability to modify and manipulate our surroundings. The technologies we now have at our disposal have amplified this potential so much that we can now create almost anything we dream of. Our knowledge of the inner realms, however, has developed much more slowly. We are probably as prone to the failings of a limited sense of self as were people two thousand years ago. This is the source of our problems. Advanced technology may have amplified our capacity to control our environment, but it has also amplified the shortcomings of our partially developed consciousness. Driven by the dictates of a derived identity, and by our belief that inner well-being depends upon external circumstances, we have misused our newfound powers, plundering and poisoning the planet.

We have reached what Buckminster Fuller called our "final evolutionary exam." The questions before us are simple: Can we move beyond this limited mode of consciousness? Can we let go of our illusions, discover who we really are, and find the wisdom we so desperately need?

Our species is far too clever to survive without wisdom.

E. F. Schumacher

These questions face us everywhere we look. Degradation of the environment is forcing us to examine our priorities and values. Political and economic crises reveal the shortcomings of our self-centered thinking. Disillusionment with materialism implores us to ask what it is we really want. The ever-accelerating pace of change demands that we become less attached to how we think things should be. Many social problems reflect the meaninglessness inherent in the contemporary worldview. And our personal relationships are continually challenging us to move beyond fear and judgment, to love without conditions. From all directions, the message is "Wake up!"

A Spiritual Renaissance

Never before has the pressure for a spiritual renaissance been so strong; and never before have the possibilities for such a renaissance been so great.

Our choice of spiritual path is no longer limited to the tradition into which we were born. We can draw from the entire spectrum of the world's wisdom. We can learn from cultures as far apart as Tibet and Peru; from traditions as different as Buddhism, Christianity, and Shamanism; from teachings given thousands of years ago, and from contemporary adepts.

Moreover, the quality of the knowledge can be preserved in ways not possible before. In the past, as spiritual teachings were passed on from person to person, translated into different languages, and absorbed by foreign cultures, some of the teaching was inevitably misunderstood or lost, while embellishments were added. What remained was a poor rendering of the original inspiration.

Today, teachings are disseminated much more accurately and easily. We can watch videos and listen to audiotapes as we travel. We can tune in to a satellite broadcast of a seminar taking place on the other side of the planet—and record it for later viewing. We can speak directly to almost anyone, anywhere in the world. We can search the Internet and draw on the insights and realizations of countless people whom we may never meet or know. For the first time, the essence of spiritual wisdom is being made globally available.

The ultimate work of civilization is the unfolding of ever-deeper spiritual understanding.

Arnold Toynbee

Whereas people in past centuries learned largely from their own experience and from those in their immediate vicinity, we can benefit from the learning of countless others around the globe. We are cross-catalyzing each other's awakening.

A Collective Awakening

When I began exploring consciousness in the sixties, there were few books on the subject. Although Cambridge had one of the largest bookstores in Britain, books on "esoteric studies" were only to be found on one small shelf in the corner of the theology section. Three decades later, the situation is very different. There is hardly a city or large town in the West that does not have a bookstore devoted to personal development and human consciousness.[1]

The thousands of books in this field published over the past thirty years reflect the myriad insights and discoveries people are making in their personal journeys. Reading these books guides or inspires some in their own awakening, who in turn pass their discoveries on to others—perhaps in books of their own, in talks and tapes, through websites, or simply in conversations with friends and family. The more each of

---

[1] I do not mean to imply that all these books are true reflections of the timeless spiritual wisdom. If there is any field of human inquiry where a cautious skepticism is advisable, it is in the so-called "new age" arena. As with any pioneering effort, many false leads are followed and wrong turns taken in the search for truth. Great care and discernment are required to sort out the wheat from the chaff.

us matures spiritually, the more we have to offer others; and the more they mature, the more they contribute to the collective awakening.

This mutual feedback not only results in an ever greater accessibility to information and guidance on inner development; it also leads to a honing of our understanding of the essential wisdom. When I discover a teaching that resonates with my inner knowing, clarifies my understanding of the mind, or adds helpful elements to my inner practice, I quite naturally integrate it into my own thinking. This is reflected in the ideas and insights I later share with others, which may resonate with their own thinking and clarify their own understanding. We are fine-tuning each other's comprehension of the essential spiritual wisdom, drawing each other closer to a common appreciation of our inner worlds.

The more souls who resonate together
the greater the intensity of their love,
and, mirror-like, each soul reflects the other.

Dante

As we share our realizations, our various expressions of this knowledge come to sound more and more alike. At a talk I gave recently, a person asked if I was saying anything that was different from what many other people were saying. My answer was "I hope not." If I am saying something markedly different, I am probably off track.

Today we easily fall into the assumption that what is new is best. We become excited by the latest breakthroughs in physics, biology, and astronomy, and are quick to embrace medical advances and new information technologies. But when it comes to spiritual technologies, what is best is that which has been tested and validated over the eons.

Our external circumstances have changed tremendously during the course of human history, and we may have very different opinions from people in the past, but the way the mind functions has not changed. The way we become caught in our interpretations of reality, the way we identity with limited aspects of ourselves, the way our attachments and fears condition our actions, the way we create suffering for ourselves—these have not changed. Nor have the basic practices that can liberate us from these impediments. In this arena it is not new knowledge that is required, but a reformulation of the timeless wisdom in a contemporary context.

The Bridge

Buddha phrased his insights in terms appropriate to ancient India, Jesus in those of Judaism two millennia ago, and Mohammed in those of his own time and culture. Today we are rediscovering that same essential wisdom and expressing it in the language of the twenty-first century.

We live in an era dominated by science and reason. For new ideas to be accepted, they need to satisfy our rational mind and be testable. It is not enough that they should resonate with our intuition; they must also make sense within the contemporary worldview.

For several hundred years our dominant worldview has been based on the assumption that the real world is the world of space, time, and matter. This materialistic model has successfully accounted for most worldly phenomena and explained many mysteries—so well that it often appears to have ruled out the existence of God.

Astronomers have looked out into *deep space*, to the edges of the universe. Cosmologists have looked back in *deep time* to the beginning of creation. And physicists have looked down into the *deep structure* of matter, to the fundamental constituents of the cosmos. In each case they have found no evidence of a God, nor any need for God. The Universe seems to work perfectly well without divine assistance.

Thirty years ago I accepted such logic. Today, I realize the notion of God that science and I rejected was naïve and old-fashioned. When we consider the writings of great saints and sages, we do not find many claims for God being in the realm of space, time, and matter. When they talk of God—the Holy Spirit, the Divine Light, the Beloved, Yahweh, Elohim, Brahman, Buddha nature, the Being behind all Creation—they are usually referring to a profound personal experience. If we want to find God, we have to look within, into *deep mind*—a realm that Western science has yet to explore.

I believe that when we delve as fully into the nature of mind as we have into the nature of space, time, and matter, we will find consciousness to be the long-awaited bridge between science and spirit.

This may be the greatest value of the new metaparadigm. In expanding our worldview to include consciousness as fundamental to the cosmos, this new model of reality not only accounts for the anomaly of consciousness; it also revalidates the spiritual wisdom of the ages in contemporary terms, inspiring us to dedicate ourselves anew to the journey of self-discovery.

———————————

Listen, friend!
My beloved Master lives inside.

Kabir

———————————

If this new worldview becomes a personal experience—a shift in the way we perceive reality rather than just a new understanding of reality—our world would change in ways that we can hardly imagine. Five hundred years ago, Copernicus could not have foreseen the full impact of his new model of the universe. Today, we can have little appreciation of how the world might be when generations have been brought up knowing that consciousness is primary, and that each and every one of us is holy.

One thing we can say: It will be a much kinder and wiser world; a world in which it will be natural to have the compassion of St. Francis, the insights of Ramana Maharshi, and the wisdom of the Dalai Lama. Freed from many of our delusions, and from much of our fear and judgment, we will no longer cause each other unnecessary pain and suffering. Inner well-being and happiness will become the true measure of social progress.

> The day will come when, after harnessing the winds, the tides and gravitation, we shall harness for God the energies of Love. And on that day, for the second time in the history of the world, man will have discovered fire.
>
> Teilhard de Chardin

By today's standards this might sound like heaven on Earth, but isn't this what spiritual teachings have always prophesied? When we realize the errors in our thinking, let go of our attachments, transcend our limited sense of self, and discover the true nature of our being, then darkness will give way to light. We will find the salvation we have been seeking, and our hearts will be at peace.

About the Author

Peter Russell gained an honors degree in physics and experimental psychology at the University of Cambridge, England, and a postgraduate degree in computer science. He studied meditation and Eastern philosophy in India, and on his return conducted research into the neurophysiology of meditation at the University of Bristol.

As an author and lecturer, he has explored the potentials of human consciousness—integrating Eastern wisdom with the facts of Western science—and shared with audiences worldwide his discoveries and insights about the nature of consciousness, global change, and human evolution.

Peter Russell was one of the first to present personal development programs to business. Over the past twenty years, he has been a consultant to IBM, Apple, American Express, Barclays Bank, Swedish Telecom, Nike, Shell, British Petroleum, and other major corporations.

His previous books include *The TM Technique*, *The Brain Book*, *The Upanishads*, *The Global Brain Awakens*, and *Waking Up in Time*. He also created the award-winning videos *The Global Brain* and *The White Hole in Time*.

Further information on Peter Russell's work may be found on his website: www.peterrussell.com.

To order additional copies of

From Science to God

Special discounts on multiple copies

Please send me _____ copies at \$15.95 (£9.95) each.

Name _____

Address _____

City _____

State/County_____ Zip/Postcode _____

Country _____ Phone _____

E-mail _____

Book total (includes shipping within U.S. and U.K.)

    \$18.50 (or £11.50) each x _____copies: _____

Discounts for more than one copy:

    2 copies: 10%   3-4 copies: 20%

    5-9 copies: 30%   10 or more: 40%

                Less_____ % discount: _____

                        Subtotal: _____

For countries outside U.S. and U.K.,
add 10% of subtotal for additional shipping.   _____

                    Total enclosed: _____

Make check payable to **Peter Russell** and mail to:
2375 E. Tropicana Ave., #733 • Las Vegas, NV 89119 • U.S.A.
or **1 Erskine Road • London NW3 3AJ • U.K.**
Credit card orders may be made online at
www.peterrussell.com